DATA, INSTRUMENTS, AND THEORY

DATA, INSTRUMENTS, AND THEORY

A Dialectical Approach to Understanding Science

ROBERT JOHN ACKERMANN

PRINCETON UNIVERSITY PRESS

Published by Princeton University Press, 41 William Street,
Princeton, New Jersey 08540
In the United Kingdom:
Princeton University Press, Guildford, Surrey

Library of Congress Cataloging in Publication Data will be
found on the last printed page of this book

ISBN 0-691-07296-5

Publication of this book has been aided by
the Whitney Darrow Fund of Princeton University Press

This book has been composed in Linotron Caledonia

Clothbound editions of Princeton University Press books
are printed on acid-free paper, and binding materials are
chosen for strength and durability

Printed in the United States of America
by Princeton University Press, Princeton, New Jersey

For

ROBERT, CARL, and ILSE

Each My Favorite

CONTENTS

PREFACE ix

CHAPTER 1

Logic and Science 3

Epistemology and Science 3

The Closure of Rationalism 10

The Closure of Empiricism 16

A Dynamic Approach 27

CHAPTER 2

Social Structure in Science 35

Social Norms in Science 35

Cognitive Norms in Science 42

Scientific Disciplines 52

Controversy and Progress in Science 62

CHAPTER 3

Science and Nonscience 74

Science and Pseudoscience 74

Science and Society 83

Science and Common Sense 96

Science and History 102

CHAPTER 4

Scientific Facts and Scientific Theories 112

The Social Construction of Scientific Fact 112

Data Domains 125

The Microprocessing of Scientific Fact 136

Theory and Experiment 149

APPENDIX

The Human Sciences 165

NOTES 187

BIBLIOGRAPHY 201

INDEX 215

PREFACE

In recent years, burgeoning sociological and historical studies of science have considerably complicated the domain in which philosophy of science might be expected to clarify scientific practice. This book attempts a synthesis of this material along with a resolution of the conflict between Kuhn and more traditional philosophical epistemologies concerning subjectivity in science. As such its aims are not modest, but it seemed to the author that in spite of the risks of failure, no attempt at a philosophical application of this material to such disputes has even been attempted. If the book belongs to an older philosophical tradition in its scope, it has had to do without the close association of logical analysis and epistemology that marked earlier efforts. Logical analysis may still accomplish much in clarifying the dialectic of fact and theory, located here as the motor of scientific history, or in developing closed axiomatic versions of scientific theories, but justificatory epistemologies and methodology have been set aside, along with associated logical functions of explanation and confirmation. Every effort has been made to minimize technicality and formalism in exposition so that the major ideas will be accessible to the widest audience of readers.

Perhaps the root failure of the methodologists was to attempt to trace scientific knowledge to the epistemological activities of the individual scientist. For reasons that will be evident in the main arguments of this book, this strategy must fail to locate a suitable sense of objectivity for scientific practice. From the classical methodologists, some echoes of Popper's fallibilism will be the major residue. Recent books by Bellone, Bloor, Latour and Woolgar, and Ravetz have influenced what follows more than citation can make evident. These books have not always been treated kindly by reviewers, but have raised issues that seem to have decisively shifted the appropriate philosophical perspective on science toward a more historical and sociological standpoint. In addition to the context provided by this background, some positions developed in the following arguments have not previously appeared in the literature on the philosophy of science.

In Chapter 1, the important insight that scientists are created through acculturation is accepted largely in its Kuhnian version. The result of this acculturation process is that scientists are always confronted in scientific activity with both theories and complex data, and are attempting to fit the two together. The contemporary philosophy of science is not, therefore, obligated to solve the general epistemological problem of priority of theory or data, but should take as its subject matter the temporal accommodation of the two. Rationalism or empiricism in the philosophy of science results when theory is thought primarily to guide data, or the reverse. Since both forms of guidance occur, rationalism and empiricism are true of episodes in scientific history, but neither can be a complete characterization of scientific progress. We must give up the idea that new theory destroys older theory, or that new theory must be an extension of older theory. The conclusion is thus that general methodologies based on empiricism or rationalism are inadequate to the dynamic interaction of theory and fact in science.

The dynamic interaction of theory and fact is then studied through other resources. The discussion introduces the view that scientific instruments break the connection between theory and observation, allowing the dialectic of theory and data to take place, and that the use of instruments establishes *data domains*, which are what theories adapt to. Instrumental splitting of theory and data means that data can be gathered independently of current theorizing, and used to constrain that theorizing. When new data domains are established by the use of new instruments, older data domains and the theories that are adapted to them do not disappear, but they can be split off as settled in principle. As instruments are improved, a succession of data domains can be used to define an objective direction of scientific progress. Another important new idea is that the acculturation process introduces diversity and individuality into scientific thinking. Part of the older context of discovery is now seen as an activity in which the sheer chosen diversity of scientific activity contributes to scientific progress, much as adaptive biological forms are selected out of a variety of genotypic possibilities. Chapters 2 and 3 develop the consequences of these ideas. A philosophy of science that has no general coercive methodology apart from the constraints of local data and tradition can find objective scientific claims emerging from diversity and selection.

Chapter 2 examines the social and cognitive norms present in science, using the results of recent work in the sociology of science. The differences between the scientific disciplines are also noted to the extent that sociological techniques can make them relevant. This investigation shows the splitting of the larger domains to which theories adapt. The existence of controversy in science is examined to see how it can select superior views without such bitter acrimony that cooperative research would be impossible. The social mechanisms of Chapter

2 are the means whereby a consensual fit of theory and data can be won from a welter of individual perspectives, a process that turns discovery into justified science.

The continual adjustment of theory to data means that successful science may not be recognized immediately, but may appear only after false starts and misleading development. Tinkering with theory and data means that scientific knowledge must emerge from this tinkering in terms of our model, and that scientific knowledge isn't (typically) instantaneously recognizable. Science differs from other forms of human activity in that its intellectual constructs and instruments are the products of a special internal evolution, but the attempt to fit resultant theory and data is structurally much like trial and error elsewhere. The view of history required for understanding scientific progress is then worked out, with some consequences for the traditional discussion of demarcation.

Chapter 4 is the presentation of the structure of science obtained from the synthesis of the ideas presented here with the valuable insights of the past philosophies of science. The interaction of theory and fact can result only when facts are seen to be negotiated data that have survived controversy, and not as a means of veridical epistemological access to the real world. Facts appear only as data text requiring interpretation. This means that fact and theory may both give way in the dialectic of progress, facts becoming irrelevant to new domains, and theories becoming maladaptive to the data domains for which they were intended. The process of accommodation of theory and fact is then studied in a series of scientific examples. Objectivity is rescued as the relationship of best explanatory theories to their intended data domains. It is suggested that much of the formal work produced in empiricist philosophy of science can contribute to an understanding of the structure of the public domain of fitting negotiated fact with adaptive theory.

In the Appendix, some lessons are drawn about the human sciences from our discussion. The same structure of accommodation of theory to fact is to be expected, but there are some interesting differences. The data domain of human science theorizing is typically either the splinter domain of the rational, utility-maximizing economic agent or the entire complex domain of man in a historical social setting. These two approaches are obtained by different points of entry of understanding into the human sciences, and can't be resolved as they stand. Attempts at progress by proposing intermediate data domains have been too theoretical, and have been rejected by the disputants because they cannot be tied down to an appropriate instrumentarium.

These inconclusive debates are examined as confirmation of the analysis of the relationship of the natural sciences and the human sciences that follows from a consideration of data domains.

The author would like to thank Donald T. Campbell, William C. Wimsatt, and Howard Darmstadter for admitting in public to having read his earlier book on the philosophy of science. Current efforts would not have been possible without encouragement from and conversation with Robert Paul Wolff, Bill Heintzelmann, Bob Saltz, Christina Erneling, Alf Bång, Bob Yamashita, as well as the euphoria consequent to the victory of the transworld depraved philosophers.

Amherst, Massachusetts
April 19, 1983

DATA, INSTRUMENTS, AND THEORY

There is no event in the history of science, not even the least that exists, such that even the most ingenious theorist can ever arrive at a complete understanding of it. . . .

— GALILEO

· 1 ·

LOGIC AND SCIENCE

Epistemology and Science

Epistemologists can begin the task of analyzing knowledge from a variety of viewpoints. At a very abstract level, an epistemologist may attempt to prove the mere possibility of knowledge in an effort to confront skepticism. This effort founders on the problem that the path of proof is not clear unless at least some matters are assumed to be known and settled. Skepticism of a sufficiently brute variety is thus difficult to dislodge by a frontal assault that does not assume as known some of the matters that a resourceful skeptic will want to keep in epistemological abeyance. Perhaps because of this, epistemologists with more to do than engage skepticism philosophically have frequently started with examples of knowledge, and have then attempted to outline the total scope of human knowledge by close scrutiny of these paradigmatic cases. A variety of examples has served in philosophical history—mathematical examples, theological examples, introspective examples, and, in modern times, scientific examples. For many modern epistemologists, epistemology is consequently the anatomy of science. Such epistemologists take scientific knowledge, or at least examples of it, to be paradigmatic cases of human knowledge. The philosophical task is then to analzye these cases to mark out the epistemologically legitimate scope of scientific knowledge.

To see more clearly what is at stake in the approach of epistemology through examples, we can turn to Plato. Plato appropriately chose mathematical examples as his paradigms of knowledge, and then sought to find the limits of the human knowledge that could be construed as similar in nature to the mathematical knowledge that he started with. There is no reason why human knowledge should be uniformly subject to the same philosophical analysis, but Plato proceeded as though an analysis of knowledge should be the same for all types of knowledge, and most philosophers have followed him in this. Philosophers of modern science often assume in a similar vein that all genuine human knowledge must be similar to scientific knowledge. Although Plato took mathematical examples as his starting point, the wisdom of his selection was confirmed by his analysis of these examples. Plato believed that genuine knowledge must be timeless and not subject to later refutation. Mathematical knowledge, as he analyzed it, was about unchanging forms or ideas, whose permanence ensured these properties. Plato then proceeded to look for knowledge in such diverse areas as ethics, aesthetics, and political theory, and he proceeded by attempting to locate forms or ideas in these areas that were similar in nature to mathematical forms.

The philosophy of science is heir to the tradition of epistemology based on examples. Philosophers of science have assumed that at least some examples of scientific knowledge are genuine examples of human knowledge, and that the way to proceed is to analyze the implications of these examples for the total range of human knowledge. This much will not be contested here as a reasonable procedure. Other aspects of the legacy are more troubling. If Plato was correct that mathematical knowledge once gained could not be refuted, is scientific knowledge to be measured against the same standard? There is the awkward fact that much of what scientists believe at any point is later found to be modified or abandoned, although the assumption that scientific knowledge continuously progresses gives hope that there is an accumulating core of scientific experimental fact that will explain this as progress.

Considered at a point in time, a scientist will be acquainted with certain supposed facts and theories, and may intend to work with these materials in order to extend scientific knowledge. An epistemology should discuss the procedure of obtaining and extending knowledge. For this purpose, the concept of rationality plays a central role in philosophical epistemology. The rational agent (or scientist) is said to find and extend knowledge by examining these materials and drawing inferences from them through the use of reason. The rational agent will respond to new facts as they are acquired, and will change the shape of his or her relevant beliefs about the significance of known facts and theories in the light of new facts only on the basis of some coherent strategy. Various facts and theoretical conjectures may be in one's personal belief structure at any time, from which basis reason will project new anticipated data, while also looking for the best possible explanations and justifications for the shape and nature of the acquired basis. Philosophy may also expect of rationality that reason will perform a policeman's role, checking that the basis be logically consistent before projection of new data. Since a person might be rational in this technical sense, but be intuitively irrational because only an excessively narrow, eccentric, and personally tailored basis had been sought, rationality may also call for the basis to be as wide as possible in terms of contained fact, and to be anchored firmly in the real world.

For all of the initial plausibility of this variant of rationality, it has some serious problems. An important consequence of this concept of rationality, even when it is fully spelled out in some particular variant, is that it is not sufficiently clear to entail the actions of a rational scientist in many of the situations that arise in scientific practice. The irritating and apparently insurmountable results of undecidability and incompleteness of modern logic suggest that interesting bases for scientific practice may not be provably consistent, and logic can press for revision only after the actual crime of inconsistency has been observed. There are in addition to these matters many problems with no known mathematical or logical solution. A salesperson with a list of towns to visit who wishes to visit each town once only on a projected sales trip, and is subject to various other constraints, cannot in general reason to an optimal route, but must proceed by instinct, or trial and error.[1] In addition to all of these problems, the consequences of much of the work that has been done on inductive inference suggest that logic cannot coerce toward a single possibility the direction in which projection from a basis to anticipated future data should take place.[2] None of these problems begins to deal with empirical uncertainty in the data, or other difficulties with the supposition that a good basis for scientific practice can be provided, but they are sufficient to show why the normative thrust of philosophies of science based on rationality is blunted by confrontation with actual scientific practice.

Similar objections stand in the path of any attempt to utilize game theoretic rationality as definitive of sound scientific practice. Where the value of a set of alternative actions can be expressed in terms of utilities, and a probability measure for the likelihood of obtaining these values on the given actions is available, rationality seems to compel the view that that end should be sought promising the highest expected utility on these measures, or one of those ends with the highest expected utility if there are more than one. This view relaxes the concept of philosophical rationality, since different probability measures can be defended by different scientists, leading to the conclusion that alternative strategies on some basis may be equally reasonable. Hope then arises that the formal philosophy of science might be brought into closer conformity with the observed practice of science through a cooperative game theoretic analysis. But problems again mount as the details are considered. Game theory is of great value and can lead to deep explanatory insights when utilities and their attainment are related in a clear way, as in many standard gambling situations. In science, as Levi and others have shown, to apply game theoretic analyses requires some notion of epistemic utilities and risks that are difficult to quantify as probabilities.[3] Where there are difficulties in quantification, as Newcomb's puzzle indicates, pairs of seemingly obvious maximal strategies may come into conflict even when the quantification is not at issue.[4] Perhaps game theoretic analyses of scientific behavior will be insightful in limited cases, where the possible outcome of a range of experiments is pretty well settled by past experience, or where a group of scientists share a sufficient range of opinion to make the probabilities and properties of outcomes of various options a matter of agreement. Even in such cases, however, game theoretic analyses would reveal similarities between scientific practice and other game strategies, and although such similarities might be illuminating of the actions of scientists in certain situations, they could not by themselves give an insight into the distinctive nature of scientific knowledge, one of the presumed objects of philosophical analysis.

All of these attempted analyses of rationality have in common the Cartesian assumption that the isolated scientist dealing with a basis for scientific practice by means of reason is the appropriate basic framework for understanding the nature and scope of scientific knowledge.[5] Philosophers may slide into this assumption by taking philosophical reflection as a model for scientific thought, or by thinking in terms of Cartesian epistemology. Its consequence, however, is a conception of scientific activity in which scientists are viewed as a group of individuals all more or less trying to do the same thing, with variance explained by the fact that some of them do it faster, or do it better, than the rest. On the Cartesian assumption that the individual scientist is the appropriate locus of philosophical reflection into the nature and scope of scientific practice, the only significance of having groups of scientists involved in scientific practice can be that the practice is speeded up, accelerating the rate of scientific progress. This is implicit in Popper's metaphor that the community of scientists can be compared to masons working on a cathedral.[6] Here the masons are conceived of as more or less interchangeable in function, and the consequence of more or fewer masons seems simply that a greater or lesser number of stones will be laid in a fixed period of time. It is the first intention of this text to challenge this natural philosophical starting point. Our considerations in later chapters will lead us to argue that the social structure of science requires for maximal advance of knowledge that the rational scientist react not only to the known theories and evidence but also to what other scientists are doing, and can do, in connection with possible research topics. Because of this, philosophical epistemologies that attempt to isolate scientific rationality by using merely the theory and data available to a scientist at a time, and logical functions over this basis, cannot achieve a satisfactory epistemology for science.

In line with prevalent philosophical practice, we have taken a basis for an analysis of scientific practice to be a personal belief structure containing facts and theoretical conjectures. Refining this conception of a basis, we can note that various philosophers have taken differing positions on the kind of facts that such a basis should contain, and on the nature of permitted theoretical conjectures over such a basis. It is also possible for philosophers to conceive of the faculty of reason used to test and enlarge the basis in different ways. The resulting complexity of possible positions is enormous, but we shall make a first division of epistemologies constructed on personal belief structures into two traditions, the empiricist and rationalist traditions.

Both the rationalist and empiricist traditions depend on a presupposition that a single person, isolated from his or her fellows, could in principle encounter the real world and reason about it in terms of ideas. Empiricists emphasize the reliability of careful sense experience as the foundation and evaluation of scientific knowledge. Empiricism as a philosophy of science has attempted to analyze scientific knowledge as the logical organization and augmentation of scientific observations. An empiricist who believes that the relevant logical organization is capable of expression within modern symbolic logic is usually a positivist as well. By contrast, one could define rationalism as the view that knowledge about the world is the development of what, in some sense, we already know in the form of clear, distinct, and mutually consistent ideas present to our consciousness. Rationalists emphasize the power of reasoned theorizing in science, and make theoretical coherence the foundation of scientific knowledge and the valuator of scientific data.

Empiricists tend not to find necessary connections between sense experiences. One experience might always be imagined as followed by almost any other experience so far as the logical analysis of experience is concerned. Logic has no empirical content, but merely allows sense experiences to be organized in a useful way. Such organization can never lead to a substantive extension of what has been observed except in the form of a hypothesis or conjecture. For rationalists, the fixed connections between ideas that are discovered by reason may be discovered in the world that is accessible to experience, but only after the connections have been thought. These connections can be recognized in the world, but they cannot be first noticed in the confusion of experience. The precise relationships of ideas are never precisely equivalent to the loose and indeterminate relationships of data.

Empiricism and rationalism have been discussed as philosophical temptations to ground scientific knowledge, and by extension all of human knowledge, in either experience or ideas. Scientists have been subject to the same conflicting temptations. Many scientists, notably Newton and many later Newtonians, have wanted the arbiter of science to be experience, and have held that mathematics and theorizing cannot discover knowledge, although theory may anticipate knowledge where it manages to accurately mirror empirical data. Empiricism is the natural ally of all experimentalists who feel that scientific knowledge can only properly advance on the basis of careful, repeated experimental results. For these scientists, some mathematical systems will be helpful in organizing data and expressing conjectures, but such systems could not be provably insightful into reality apart from their shaping in experience. Experience has to perform the separation between what is useful and what is not useful. Other scientists have seen experience as a snare, partly because of errors of perception and measurement, and have seen experience as little more than a check on whether the precise assumptions of theory point in the right direction. For these scientists, reality must have the precision and beauty of a mathematical development if it is to be understood, and hence mathematical system must be the primary instrument for insight into natural process. Only mathematical systems can transcend the limitations of human perception.[7] Rationalism is the natural ally of all scientists who see theorizing as the primary motor of scientific advance. Scientists have generally spent little time on the philosophical articulation of these themes, but it should not be imagined that the incapacities and problems of empiricism and rationalism are totally isolated from tendencies in the practice of scientists.

Empiricism and rationalism, however, have joint difficulties that cannot be traced to the clash between their intuitions about the relative importance of theory and data. Neither empiricism nor rationalism can satisfactorily explain sufficiently *rapid* theoretical changes in the sciences. Both have historically seen the object of knowledge as unaffected by the process of acquiring knowledge. After the knower studies the object of knowledge, it remains unaffected, and the knower changes primarily by adding information about the object to the stock of his scientific basis. These epistemologies would be quite serviceable if the world were stable, and if the main features of the world could be read by human sense organs and human reason. There is a sense in which these epistemologies were defensible before the accelerated growth of modern science and the confusion of data produced by modern scientific instruments. If the properties of the world as revealed in experiment change sufficiently rapidly, then the relatively static traditional epistemologies will not prove adequate to science, since the fit between language and data on which they depend will be constantly disrupted by newer data produced by a rapidly changing instrumentarium. It will be argued here that the features of the world revealed to experiment cannot be philosophically proven to be revealing of the world's real properties, but that experiment produces a text of data that must be interpreted, and whose augmentation may not seem initially consistent. The complexity of the world revealed by modern science, as well as the failure of epistemological independence between knower and known (and between theory and data), seems to point inevitably toward a newer epistemology if the structure of science is to be captured at a satisfactory level of philosophical analysis.

The case against rationalism and empiricism need not rest on any claim that they are *prima facie* inadequate to current scientific theorizing based on intuitions about the dynamics of scientific progress. In order to motivate serious consideration of the more dynamic epistemology to be presented in this book, it will be argued that both empiricism and rationalism suffer serious internal difficulties. Neither the idea that knowledge can be grounded in clear and distinct ideas nor the idea that knowledge about the world can be developed as properly organized clear sense perception can result in a self-consistent epistemological position, that is, a position that can legitimate its own view of scientific knowledge. This is the root problem with these traditions and their informal allies in scientific practice, although the consequences only prove devastating where data development is extremely rapid. Rationalists are unable to establish that knowledge read from clear and distinct ideas is knowledge about the same world studied in modern scientific experimentation without assuming the legitimacy of at least some sense experiences, that is, without accepting some minimal empiricist assumptions. Empiricists are unable to establish that knowledge about the world exceeding the knowledge represented by past sense experience can be obtained by logical means without assuming the legitimacy of at least some intuitive idea about how the world is structured, that is, without accepting some minimal rationalist assumption. Neither epistemology, therefore, can be satisfactorily closed in terms of its own assumptions.

Both rationalism and empiricism, as well as their scientific allies, are partly correct in their insights. At times, scientific advance is powered primarily by theorizing or by experimenting, and on such occasions the relevant epistemological position is completely serviceable in practice. When a scientific language suitably reflects the facts of experimentation over a period of time, either rationalism or empiricism may seem correct and defensible. Both make assumptions that are sometimes reasonable in a scientific context, and this helps to explain why variations of these classic positions tend to recur in writing about the epistemology of science, and why particular scientists can see their work as consistent with one view or the other. In attacking each as a partial or incomplete account, we also need to preserve the important moment of insight that accounts for its longevity. Then after this comparison, it will prove possible to expose the apparent clash between them as mistaken, and to find a more dynamic perspective within which the advantages of both can be preserved.

The Closure of Rationalism

It has seemed so obvious to so many philosophers that modern scientific knowledge is grounded in the close observation of nature represented in experimental results that some form of empiricism underlies many contemporary works on the philosophy of science. The implicit challenge to the tradition of rationalism in connection with modern science is to show how one can, by contrast, read reliable knowledge from one's ideas. Since our ideas are frequently contradictory or unclear, rationalism must restrict the ideas from which reliable knowledge about reality can be read to carefully prepared candidates for a theoretical system. Turning to an example of how ideas can apparently augment knowledge, we can consider the transitivity of the relation *longer than*. Clearly, if one suitable object is longer than a second, and the second longer than a third, we feel sure without further measurement that the first is longer than the third. Suitability here rules out change in length, and includes a clear significance for the direction of length. Experimenting on objects to determine whether this relation is transitive seems pointless, and the transitivity is consequently not merely a contingent conjecture. Such a principle is neither formally provable nor logically valid, and yet it is not supported merely by fact.[8] Some philosophers have said that such a principle is true because of the meanings of the terms involved. Should this be so, the unobservability of meanings seems to force the view that this principle is a paradigmatic case of rationalistic knowledge.

Another example will indicate how useful scientific information about the world seems to flow from a principle based on ideas plus thought experiment. We will call this the Brake Principle. The Brake Principle can be formulated as follows: Two bodies moving in parallel paths at different velocities will, if then rigidly coupled, not be able to move in coupled form at a velocity greater than the faster of the two velocities before coupling. To put this in quite informal terms, a slower body will typically act as a brake on a faster body to which it is rigidly coupled, and can certainly not speed it up. Two horses or two humans coupled together cannot presumably run more quickly coupled together than the faster of them can run alone. Thus, thought may convince us that the Brake Principle may be used legitimately in physical reasoning. Suppose we now consider the rate of free fall of bodies in the earth's gravitational field, once a significant unsolved problem in physics. Let us consider a batch of objects, all of the same size and weight, manufactured so that two, three, or more of these objects can be rigidly fastened together. We can imagine, for example, that the objects are threaded metal cylinders that can be screwed together in arbitrary combinations. Suppose each single cylinder weighs one pound. We can imagine dropping simultaneously two objects, one of them a single cylinder and the other three cylinders screwed together. There are three possibilities. The three-pound object can fall faster than the one-pound object, slower, or at the same rate of speed. The Brake Principle is sufficient to rule out two of these possibilities, leaving us with a significant fact about the world.

Suppose, for example, that the three-pound weight falls faster because heavier bodies fall faster. If we couple the two objects together to get a four-pound object, it should fall faster then the three-pound weight by this line of reasoning; but it can't fall faster according to the Brake Principle, since the four-pound weight is the rigid coupling of the one-pound and three-pound cylinders. This contradiction eliminates one option. A similar line of reasoning is sufficient to find a contradiction between the assumption that heavier bodies fall more slowly than lighter bodies and the Brake Principle. But if all bodies fall freely at the same rate of speed, no contradiction appears.[9] These results hold when such factors as air resistance are considered, and discounted, so that free fall in the gravitational field is made as precise as one wishes conceptually. The Brake Principle in conjunction with this thought experiment seems to entail that all bodies freely falling in the earth's gravitational field must fall at the same rate of speed, a physically significant fact. The necessity involved follows from the contradiction entailed by any deviation from this point of view.

No actual experiment has been performed to arrive at this conclusion, and once this line of reasoning has been discovered, an experiment may seem superfluous save as a crude check that no major error in thinking had occurred. Many readers will suspect that the straightforward attack on this problem is to drop pairs of objects from a height. From a rationalist point of view, however, observation cannot determine whether objects are dropped simultaneously, or whether they hit the ground at the same time. An observation could determine that the objects hit more than a second apart, but observation can't determine whether two objects hit simultaneously or a thousandth of a second apart, a difference that may be crucial to theory. It is assumed here that naked eye observation is the relevant means of determination. Experiment can show that we are not wildly wrong, but can it find the truth about nature? Many hypotheses are always logically compatible with experimental observation. Only reasoning based on correct ideas can seem to penetrate this confusion and locate scientific truth.

We hope this example is sufficient to show the often unrecognized power of reason. The ideal rationalist, as gentleman, is prepared to wait for these moments of convincing insight, no matter how fruitless a temporary search for results. Rationalists will not spread mere rumors about nature, a lady. Empiricists, who are prepared to violate nature through experiment, can force their advances on her, but she will not necessarily tell them the truth. It is undeniable that there is a rationalist component in much of science, and in nearly all theorizing. Many areas of science have received extensive and persuasive rationalist organization, with experiment merely seeming to corroborate a structure of ideas that can literally stand on its own. And theories about ideal gases, frictionless surfaces, perfect competition, and so on, seem to tell us something about the world even though no experiments can be performed directly on these objects. The world of scientific theorizing is a world of precise intellectual constructs. Rationalism must locate enough such structures to provide a suitable account of contemporary science and of the dynamics of scientific progress, but it falls short.

Where language has been sufficiently often used successfully in interacting with the world, and perhaps even modified to produce this success, the ideas it can express may seem to reveal the precise structure of the world. But by examining ideas alone, even clear and distinct ideas, we can discern no relationship between these and the empirical world open to sense experience that philosophically legitimizes rationalist knowledge as knowledge about this world. One cannot postulate a correspondence of some kind, because insofar as one cannot get away from ideas to examine the world directly, one cannot externally examine a relationship between an idea and its referent in the world. Such a relationship transcends rationalist knowledge, since within rationalism the only relationships are those between ideas. Much of scientific knowledge, of course, can be seen as a set of relationships between theoretical ideas and conceptualized data, but this still omits any legitimation that the conceptualized data are reflective of the actual world. Perhaps a more promising direction is to believe that evolutionary adaptation or God's benevolence has resulted in a correspondence between ideas and the relevant objects in the world. The attempt to reach beyond the circle of ideas to a correspondence with the world through God's benevolence, such as the attempt in Descartes' *Meditations*, has never resulted in an argument demonstrating more than the possibility that such connections could exist. Any evolutionary argument has similar problems, and also has problems with the rate of possible theoretical development. Perhaps ideas have evolved so as to correspond with important features of the world, but changes in the world, or in the world of data, could leave such ideas without purchase. Indeed, in this respect, the difference between changes in the world and changes in the data that we use to describe the world is moot. What is required is a method of forcing ideas to develop so as to accomodate empirical evidence in such a fashion that new discoveries can be reasonably anticipated. This strategy cannot be accommodated within traditional rationalism, since an empirical input is required as a guide, no matter how uncertain, for the direction in which ideas must be developed. A self-consistent rationalism comes close to the vacuous assertion that empirical knowledge can be read from ideas whose structure suitably mirrors the structure of the world, without telling us how such ideas can be recognized and acquired.

Both phenomenology and Platonistic philosophies remain viable in spite of this critique, since they do not purport to give us knowledge about the empirical world, but about at least some of our mental acts and about ideal mathematical objects, respectively. This is perhaps why they are still viable forms of rationalism in special subject matter domains. Although not subject to an internal critique, these forms of rationalism may be subject to external critique. As is well known Husserl was never able to extend phenomenology to an account of societal interaction between human beings, and modern forms of phenomenological sociology seem to be heir to this original difficulty.[10] In this sense, pure phenomenology is insufficient to explore the entire human world. Platonistic philosophies of mathematics would be open to a charge of superfluousness if philosophies of mathematics could be worked out to account for the mathematical knowledge we need for science and other endeavors that do not need to postulate independently existing mathematical ideas. Perhaps the necessity of mathematical truth can be captured in the rules of constructivist proof, or in some notion of reflected social structure, but the debates over these questions are anything but settled. In view of the fact that pure logical structures do not seem adequate to the development of mathematics, some version of nonempiricist mathematical truth still seems an attractive possibility, and at this point it is impossible to rule out. What we can say here is that modern rationalism avoids entanglement with dialectical problems through restriction of its claims to truth to the relatively settled underpinnings of everyday judgment or to mathematical truth.

As has been noted, there is a pressure in rationalist thought toward comprehensive system, with the idea that a sufficiently vast, consistent, and coherent account of the world would have to be true. In such a program, revision to bring about consistency might cause any part of the system to require repair. In this view, specific ideas need only be generally related to empirical reality, since the ultimate fit of system and reality could only be assured when the system was sufficiently complex. Rationalism thus has some difficulty in accepting the structure of the sciences, which according to the scientific division of labor allow for small domains of fact to be closed off and definitively settled from the perspective of theory.

The rationalist tradition has observed correctly that one can read knowledge from ideas, including ideas that have been expressed in linguistic systems. But knowledge about the world, scientific knowledge, can only be read from ideas or language that happen to fit the world or isolated domains of fact within the world. Then the fact that the fit exists need not be kept constantly in mind, and the role of past experience in developing such a fit can be forgotten. As our Cartesian example shows, our everyday language fits the world fairly well as the result of a long history of its adaptation to human observations and human interests. The fact that we experience objects of stable size that we can easily reidentify in much of everyday life forms part of the essential background in which the transitivity of tallness seems essentially correct. Similarly common experiences with moving objects of normal dimension and speed are part of the background of the plausibility of the Brake Principle. A great deal of knowledge about the world has already been coded by an evolutionary process into our vocabulary and the way in which we use language. Early scientists required little more than this everyday language to describe the world and to speculate about it, and experimentation did not always need to play a big role in scientific advance. It was possible then to be optimistic about the completion of a rationalisitc program and the development of a language that could completely and accurately describe the world. Crushing problems for traditional comprehensive rationalism arose with the rapid advance of modern science. The data obtained from many new scientific instruments, especially in the natural sciences, are not naturally dealt with in many cases in existing languages, and they may seem alien to expectations based on past experience and present languages. Scientists have had to develop their own languages to describe such data and to pursue theoretical development. But in this process the basic insight of rationalism is not lost. Languages are sought that will fit data so closely that information can be read from proper linguistic expression without further experimentation, and the history of science shows that such languages have been repeatedly developed.

Chemical symbolism provides a good example. Using it, a chemist can anticipate the success of chemical experiments, which therefore need not actually be performed, as well as the impossibility of hypothetical compounds, whose synthesis therefore need not be attempted. The symbolism of chemistry was originally used to mark the proportions of reagents involved in certain experiments, but it came to fit chemical structure so closely in certain data domains that chemists could take it to represent the actual structure of the molecules whose study constitutes their science.

There undoubtedly is a strain of rationalism in science. Scientists seek to develop languages like that of chemical symbolism to facilitate the growth of science. Languages with sufficient fit to the available data and to potential experience make a suitable instrument for guiding scientific advance. The development of such languages cannot be explained save partly in terms of language that is already understood, language that has earned trust in terms of its wide consonance with the structure and process of the experienced world. The failure of rationalism in the context of modern science is its failure to explain how language and the world can be brought into consonance when the experimental interaction with the world is producing data that break the confines of existent expression. Rationalism must assume that existent expression is adequate. We need to understand the role of scientific inquiry in forming legitimate semantics and syntax for scientific languages. Rationalism is correct that the existence of these languages is crucial for science, but it had precious little to say about the process by which adequate languages can be formed.

The Closure of Empiricism

We will consider the empiricist tradition primarily in the modern form most frequently encountered in the study of science, that of positivism and consequential positions. This form of empiricism presents the issues that concern us directly, since even if it be conceded that positivism is self-consistent, its failure to provide a complete grounding of scientific knowledge solely in experience will show convincingly that empiricism cannot yield an adequate account of science. Positivism is a form of empiricism that arose after the development of modern science explicitly to explain its epistemology. The failure of positivism is partly the failure of one form of scientific self-understanding.

The first positivists in the 1920s were concerned to argue that science could be sharply distinguished from theology, common sense, and even logic and mathematics. In this way they hoped to show that science was sharply superior as a means of acquiring empirical knowledge. In order to draw a boundary around science to highlight its unique status, early positivists construed science as resting on paradigmatically clear, carefully prepared observation and experiment, and they tended consequently to downplay the role of imagination in theorizing. Creative theorizing, wherever it occurred in science, had to be subject to definite and vigorous control through scientific observation. In this way theorizing in science could be distinguished from theorizing in other areas of human interest. Although logic and mathematics were construed as having no substantive content, they were useful in organizing scientific knowledge and in legitimating inferences from accepted observation statements and sets of such statements.

Positivists were anxious to sweep away theology, aesthetics, and so on, as intellectual rubbish, and to restrict knowledge to scientific knowledge. There is in this approach a kernel of the booster, and positivists often sounded like pitchmen even if the results of science they were interested in were real enough. Some positivists were anxious to draw the line around science so as to include sociology as a science, while others thought that perhaps only physics, chemistry, and biology belonged clearly to the sciences. We will not concern ourselves with the details, but with the shape of the common structure of science admitted by the positivists. As we have already noted, this structure was designed to show that science represents the most certain knowledge available to human beings, logic and mathematics being regarded as auxiliaries without content, and a form of knowledge that might be indefinitely expanded through controlled experiment and research.

The general features of science, as understood by the positivists, were roughly the following. First, science is a clearly demarcated body of knowledge that is the only worthwhile object of philosophical contemplation. The clear line of demarcation between science and non-science is roughly (or even exactly) equivalent to the boundary between sense and non-sense, or rationality and irrationality. Second, science is the only hope for the future of mankind, the only instrumentality for successfully solving the problems facing the human population, and as such deserves the support of philosophers. Third, science has a tripartite structure consisting of *logic and mathematics*, which provide the language within which scientific results are formulated; an experientially given *basis*, the factual content of science; and at least some *theories and hypotheses*, which could provide rational expectations about the future given the experiential basis. Fourth, the future of science is optimistic. Since the experiential basis can be indefinitely expanded through controlled experiment, enough information could, in principle, always be accumulated to provide the answer to any intelligible questions. Fifth, there is a potential division of labor between scientists and philosophers. Scientists are constantly expanding the experiential basis and proposing theories, but philosophers have the expertise to decompose and analyze the relationship between the basis and theories proposed in order to determine what part of the basis any given theory can be said to *explain*, and which theories are best supported or *confirmed* by the experiential basis. Even if a general characterization of explanation and confirmation is possible, the precise values of the relationships of explanation and confirmation linking the basis with available theories will require updating as the experiential part of the basis expands, and philosophers can find here a constant source of service in support of the progress achieved by their scientific colleagues.

The major internal tension in positivism is caused by a combination of the first and third characteristics cited above. According to the first characteristic, storytelling (including metaphysical stories) is to be banned from science as not sufficiently grounded in the experiential basis of science, but the problem is to include scientific theories along the lines suggested in the third characteristic, some of which are pretty wild, imaginative extrapolations from accepted data, while ruling out nonscientific storytelling as having no cognitive purchase. The famous rejection of metaphysics was required if positivists were to feel that they knew exactly what was being discussed among scientists so that they could map out the appropriate logical relationships. It seems dubious whether scientific theorizing can be accomodated within such a conservative outlook. No one has ever found an analytically precise way to include scientific theories and hypotheses within science, excluding other forms of storytelling and fitting scientific intuition about the extent of proper theorizing.[11] Positivism has provided a restrictive account of theorizing in science because of one attempt to restrict the meaning of theories to some close logical tie with the data in any basis. Some positivists have simply accepted the consequences of their analysis as binding on science, and they have drawn the conclusion that our normal intuitions about the role of theories and hypotheses in science are quite extravagant, if not incorrect.[12] While this illustrates one of the charms of philosophy, we shall pursue here a path designed to protect the intuition that the use of theory and the importance of constantly developing new theories are essential components of scientific progress.

This brief summary of positivism must inevitably brush aside some subtleties that have been at the center of positivism's own justifications for the relationship of philosophy and science that positivism wishes to defend. In the logical analysis of science, positivism brings to bear on scientific practice a normative logical yardstick that can evaluate practice in terms of logical rigor, an evaluation that must ultimately complicate any crudely defined border between philosophers and scientists. This normative purchase, of which positivism can rightly be proud, is leached away in any philosophy of science that makes total scientific practice the sole standard of scientific rationality. Many positivists also drew on a distinction between the context of justification and the context of discovery, arguing that their logical methods could be realized only in the context of justification, that is, *after* sufficiently sharp statements of theory and data were available for logical models to come into play. In the heuristics of theory proposal, however, scientists were free to draw on any sources, no matter how irrational they might seem. Thinking of a theory and then stating it for other scientists and confirming it were regarded as two quite distinct steps. The process of discovering new theories was conflated with an unanalyzable, and possibly irrational, sequence of acts. The process of confirmation, on the other hand, was subject to a rational and public corpus of rules for proceeding with the analysis of the degree of support of the theory by empirical evidence. In view of the fact that the distinction itself seems implausible and difficult to maintain, but requires extensive and subtle development if it is to be assessed, and the fact that a position will be developed here that partially analyzes theory development against the control of data domains, we shall proceed here with an evaluation based solely on positivism's analysis of the logical structure of the context of justification, which by its own admission is positivism's major contribution to the understanding of science.

We can now turn directly to the internal question of the failure of closure of positivism. It is an astonishing fact that no comprehensive and adequate account of the relationships of explanation and confirmation between the experiential basis and the layer of theory in the positivistic account of the structure of science has ever been provided, even though such relationships are frequently cited in empiricist practice. Neither the concept of explanation nor that of confirmation can be closed according to empiricism for reasons to be developed shortly; and to accomodate the observed features of scientific practice, both must be taken to involve at least some assumptions that cannot be justified solely in terms of the experiential basis and logical considerations. The logic assumed historically by the positivists to be the logic of science, roughly the first-order predicate calculus, seems inadequate to capture these important concepts in terms of the data basis. More recently, intensional, modal, many-valued, and other kinds of logical systems have been utilized in an attack on this problem, so far without success.[13] This diversity, of course, threatens the conceptual unity attempted by positivism, but we cannot rule out that some exciting developments may be in store. The appropriate relationships may one day be established at the level of data domain and theory to be developed here, but the extent of the problems with a purely empiricistic set of relationships can be developed in the context of positivism.

It seems clear that acceptable scientific theories in a basis at any time must explain at least some of the facts in the experiential basis by permitting deduction of statements representing these facts from other parts of the basis. Positivists originally took the position that deductive scientific explanation of this kind (with suitable restrictions) was the sole requisite kind of scientific explanation, but controversy has arisen on this point. The statistical inferences encountered in many branches of contemporary science may have to be explicated in terms of a distinct notion of statistical explanation if there is to be a comprehensive philosophical account of scientific explanation.[14] There may, therefore, be at least two relevant notions of explanation associated with at least some bases. The original models of deductive explanation were also restricted somewhat by inability to specify formally what constituted legitimate scientific theories, laws, and hypotheses. Because of this, no one could say which explanations or apparent explanations actually contained scientific laws, theories, or hypotheses. An explanation could be defined as a syntactic or logical relationship involving putative theories (or laws or hypotheses) and experimental statements, but a semantics could not be worked out that would eliminate only accidentally true generalizations. Empiricism thus wavered between providing an analysis of explanations and explanation sketches, the latter having only the syntax of proper explanation.[15] All of these problems can be waived, including the precise distinction between laws, hypotheses, and theories, since the problem of closure will arise for deductive explanation even if we assume that all of these niceties have been satisfactorily resolved.

The most widely defended model of deductive scientific explanation is due primarily to Hempel.[16] In this model the deductive inference that is an explanation has two premises and a conclusion. One premise must be a theory, or conjunction of theories that is used as a theory in the relevant explanatory deduction. The second premise, usually found in practice, expresses some known facts or boundary conditions that are necessary to utilize the theoretical premise in a particular case. A satisfactory scientific explanation is then a deduction of the statement to be explained from this pair of premises.

Although scientific explanations may, at least on some occasions, fit the model, and although scientists would be embarrassed to discover that some explained statement did not follow logically from suggested theory and certain background facts, it is quite clear that the Hempel model in its full generality does not capture the intuitive structure of sound scientific deductive explanation. An argument having the structure of a Hempelian explanation may simply not capture any intuitive relationship of relevance between its premises and conclusion of the kind that scientists would expect. This has been shown rigorously by Eberle, Kaplan, and Montague.[17] Thus the full deductive model does not seem to capture the intuitive relationship of relevance between an explaining theory and a factual statement that is to be explained in the general case. The upshot of this result is that the relatively clear Hempelian model has had to be replaced by a group of explanatory models, each member of which offers a restricted model of explanation that can be regarded at best as a sufficient condition for scientific explanation if the relevant semantical properties are satisfied.[18] It is absurd in this situation that some philosophers should proceed on the assumption that the empiricist notion of scientific explanation is a satisfactorily clear concept.

We have seen that the Hempelian model is too loose, that is, that it lets in at least some intuitive nonexplanations, and hence requires restriction, but it is also too stringent. Many, if not all, of the explanations accepted by working scientists in the course of normal scientific activity fail to be Hempelian explanations. The basic reason for this is that while scientists want what they explain to follow from what they already know and from the theoretical apparatus they believe to be true in the straightforward sense that the premises of an explanation couldn't all be true while the explained statement was false, they are concerned with this relation only in the space of scientifically plausible worlds, and not in the space of all logically possible worlds. The difference is important. If a critic can describe a scientifically plausible world in which an explanation's premises are true and the explained statement false, any scientist will recognize the need to attempt repair of the putative explanation. On the other hand, the fact that one can describe a logically possible world in which the premises of an explanation are true and the explained statement false need not have an impact on scientific practice. The set of logical possibilities is much wider than the set of scientific possibilities, and some logical possibilities will be scientific implausibilities.

Let us imagine an anthropologist interested in demography who suspects a migration to have occurred between village *A* and village *B* during a certain period, where the villages are part of an extensive archaeological exploration.[19] She is concerned first to prove that a migration did take place, and then to speculate on the reasons for the migration. Perhaps the demographic data, derived from archaeological findings, show the two villages to have been relatively isolated, and the villages surrounding them to have remained stable in population during the period in question. We will also assume that the demographic data indicate that the suspected migration occurred so rapidly that the shift in population from *A* to *B* could not have been the result of a natural population decline in the one village and a natural population rise in the other. While we're imagining the facts, let us grant data so extensive that any reasonable anthropologist would concede that a migration took place on the strength of the evidence. But while the fact of the migration may seem the only reasonable interpretation of the data, it is not the only logical interpretation. There is no theory that will derive the statement of the migration in a Hempelian explanation from the data, and there cannot be. In the total space of *logical* possibilities there are such possibilities as a spaceship's picking up some or all of the population at village *A*, reprogramming it, and somehow returning it to the wrong village. We can assume here that this occurred without a trace in the relevant history. Anthropologists simply do not normally consider the possibility of intervention by space visitors, which would enormously complicate their work, but our imagined data could not rule it out. Of course, if evidence *for* intervention by space visitors became accepted, anthropologists might evaluate the migration data in the wider perspective. This example is sufficient to indicate how good inferences made within the constraints of discipline plausibility may be invalid for the logician, who cannot recognize these constraints.

Scientific explanation is actually highly dependent on context, for the set of relevant possibilities to be considered at any point in time is determined by the then current state of scientific knowledge and the associated space of scientific plausibility. It would not be helpful to science to demand that logical rigor be met, since the necessary excluding clauses would make for inelegant and implausible scientific explanations, and they would have to be altered in the event of scientific progress anyway. Beyond the common-sense agreement that explanation should be deductive within the appropriate space of possibilities, the allegedly greater rigor and uniformity of logical tests is a complicating factor in the pragmatics of explanation that does not seem to have a clear compensating advantage for scientific practice. Scientists will frequently have a relatively precise agreement on the competing possibilities because of extensive discussion and negotiation, but while this may lead to more uniform patterns of inference than are encountered outside of science, it will leave logical rigor unaffected. Invalid arguments are invalid arguments, consensual agreement or not. Philosophical accounts of scientific explanation would prove more illuminating if philosophers were to study these local contextual constraints, rather than pursuing the path of promoting an excessively idealized logical paradigm with no hope of advantage for scientific practice. But to concede this point is to modify considerably the idea that general methodological remarks can be insightful in all areas of scientific research.

Let us turn to the complementary notion of support or confirmation. Initially the notion may seem equally as central as that of explanation, perhaps because scientists do discuss the evidence in favor of various theories. We can imagine that two different theories are both useful in explaining the same range of data, or roughly the same range of data, even though they can be shown to diverge on anticipated data. It is legitimate to ask which theory should be pursued, and which theory ought to be accepted for the time being in the pursuit of practical goals. If we could say that one theory was more likely to be true than the other, then we would have a reason to prefer this one for such purposes. At the same time, one might be skeptical, cautious, and rationalistic, and refuse to speak of one as more likely true than the other. In this case one could wait until the data differentiated the rival theories, and adopt an agnostic position regarding their merits in the interval.

One version of this policy is the critical rationalist stance of Popper, and it can't be faulted logically.[20] On Popper's view, one considers only falsifiable theories in science, and one must reject them only if they are falsified by the data. For practical purposes, one might adopt the most highly corroborated theory in a stable data environment for development, or for practical purposes, but the fact that the data environment might change radically and unexpectedly rules out reliance on any inductive inferences to a most likely true theory or hypothesis. Anticipating later discussion to a certain extent, we can notice at this point why confirmation is so problematic on epistemological grounds. Two scientists may be forced to agree after examination of the facts that some theory explains some range of data, even if one of them doesn't find the theory otherwise promising. Now suppose that two scientists are forced to agree after examination of the facts that one of two theories is better supported by the current data. It can hardly follow that it is irrational to pursue development of the weaker theory on suspected future evidence, and it may be helpful in locating the best theory if the two scientists divide up their labor, each working on the development of one (but not both) of the theories in a cooperative rivalry, or at least a rivalry constrained by certain norms. Because of the possible advantages of such a division of labor, it is not clear what consequences a precise empiricist theory of confirmation might have for scientific purposes.

The failure of closure of empiricism is quite clear with respect to inductive inferences to a most likely true theory or to a theory with maximal confirmation. No matter how confirmation is construed, as long as it measures in some sense the intuitive notion of support, this observation will remain correct. This will be illustrated here by an example that will exhibit the primary reason for this failure.[21] In our example, we will attempt to learn something about a coin by tossing it and observing the outcome of the tosses. This example will stand for the general problem of learning from experience. Suppose we have previously manufactured two identically appearing coins, one of which has a probability of one in two of coming up heads when tossed by a mechanism designed to randomize outcome, and one of which has a probability of three in four of coming up heads when tossed by the same mechanism. The latter coin is said to be biased toward heads.

We are attempting to learn which of the two coins we have by tossing it. Without going into mathematical details, we expect the sequence of heads and tails to have a higher probability of matching the sequence predicted by the true hypothesis than by the false hypothesis. If we toss the coin one thousand times, and we obtain approximately three in four heads or one in two heads on the tosses, we are pretty sure which coin we have. Statistical induction proceeds on such intuitive bases, even though the mathematical theory can be used to compute precise probabilities and to settle intuitively obscure alternatives. We know in theory that we can never be certain about any such inference. In the case of the coin, logically all of the tosses could show heads no matter which hypothesis was true. In spite of this fact, we have little trouble reaching seemingly settled convictions on the basis of statistical reasoning in clear circumstances, and we can easily develop the formalism that seems to express this situation.

Let's look at what we have to assume in order to legitimate our inferences from experience in the case of the coin. If we set down all of the possible specific sequences of 1,000 tosses before we start experimenting, it may seem tempting to suppose that in regarding all of them as equally likely, we are not making any empirical guess about the actual nature of the world, but are simply expressing the logical situation. Now suppose we have tossed the coin 999 times and have 749 heads among these tosses. Intuition says that the coin is the biased coin in this situation, and that heads is therefore more likely than tails on the next toss. But we have no mechanism for changing our original assumption, which is that heads and tails are equally likely on any particular toss, hence on the last toss of the coin. The original assumption thus turns out not to be neutral at all. It does not provide a logical space within which we can make discoveries. It actually legislates that the coin is fair, after which we can't discover from experience that the coin is biased. No matter what we have obtained on our 999 tosses, either heads or tails remains equally likely on the next toss.

We can try another assumption. The old assumption was that there is no connection between what we observed and what we hadn't observed, that they were independent. Let us assume in a simple way that what we haven't observed is likely to be similar to what we have observed. This is not to refute Hume, but to assume that he is wrong. We will assume, not that each specific sequence of 1,000 tosses is equally likely before we start, but that we are as likely to get any fixed number of heads in 1,000 tosses as any other. This assumption has an interesting consequence. There is only one specific way of getting 1,000 heads or of getting 1,000 tails. There are more specific ways of getting 999 heads or tails, and this number continues to rise until we look at the specific sequences for getting 500 heads or 500 tails.[22] If we assume that all of the ways of getting a fixed number of heads are equally likely, and that we are as likely to get any fixed number of heads as any other, our expectation before experiment has a different structure. Of two specific sequences that have 500 or more heads in them, and that differ in that one of them has a greater number of heads in it, that one is more likely that has more heads, because it comes from a smaller set of original possibilities, which have to divide up equally the equal probability of getting a fixed number of heads.[23] Now suppose we have 999 tosses in which there are 749 heads, and we are anticipating the last toss. It is more likely to be heads than tails, in virtue of this distribution. If we have 999 tosses in which we have 500 heads, the last toss is also more likely to be heads than tails, but there is a difference in the probabilities. The probability of getting a head on the next toss when we have 749 heads is larger than it is when we have 500 heads, since our method of deriving probabilities assigns getting 750 and getting 501 heads the same probability, but the probability of getting 750 heads is divided by fewer ways of achieving this result. Thus, we can apparently learn from experience if we make this assumption concerning the probabilities before we start. We can argue that the probability of getting a heads on the last toss is greater if we've had 749 heads than if we've had 500 heads, quite in conformity with intuition.

In our example, we have seen that we can obtain what seems the intuitively correct result by making some empirically unjustified assumptions about the nature of the world. The history of this point is a little more subtle. At first it was thought that the assignment of prior probabilities was a *logical* measure over a language describing the world. Now since different logical measures result in different prior probability assignments, and hence in different appraisals of the significance of the coin tossing, we are left with the alternatives that either logic has content or some substantive presuppositions must be made if a logic of confirmation is to produce results in practice. Since the claim that logic has no empirical content was important historically to the positivists, they were forced gradually into the position of conceding that some substantive presuppositions are necessary in order to obtain a suitable theory of confirmation. No self-consistent empiricism employing a notion of confirmation, therefore, seems to be possible.

The nonempiricist assumptions required to augment empiricism can be squeezed into a choice of formal language or into a choice of logics, but they are unavoidable. If a formal language is chosen for expressing scientific statements, it is a selection from a set of possible languages, and any logical distribution of probabilities over its simplest sentences will reflect the substantive aspects of the structure of the particular language chosen. A language that asserts that there are an infinite number of objects and one that does not so assert will lead to different probability measures that will be reflected in different confirmation estimates in at least some cases. But even if a language can be found that achieves only intuitive results, the move to a formal language is a concession to rationalism, since rationalism can be viewed as an attempt to find languages from which correct information about the world can be read in scientific practice.

In the context of everyday inference, scientists make assumptions about the world that affect their notions of evidential support. They are likely to assume that only a few explanatory possibilities exist, and discuss which of these is most likely given the evidence. These arguments may be so informal as to defeat the point of formal philosophical clarification, partly because the scientists involved see the context of these discussions as constantly altering. For example, the use of statistical inference nearly always *assumes* that some distribution is present, and then attempts to work out the consequences of that assumption. This is to bypass the philosophical problem of induction, rather than to solve it. For example, one may assume that a population is normally distributed in a standard way, and then attempt to calculate the value of the mean and the standard deviation of that population on the basis of the empirical evidence. Sampling plus the assumption about the general distribution leads to the desired description of the total population. This sort of problem transcends empiricism because the assumption about the distribution cannot always be empirically justified, and because the kind of distributions considered will depend on the plausibility space of the local scientific investigation. Scientists are just likely to try what has worked in what they see as similar situations, an attitude that flirts with profligacy for the methodological puritan. We will attempt to show how the intuitive correctness of this procedure can be rationally grounded once the constrictions of strict empiricism have been laid to one side.

If empiricism cannot be closed, why do so many philosophers remain committed to empiricist programs, attempting to refine empiricism rather than looking for alternatives? Such philosophers seem in fact to exhibit the profile of Kuhn's normal scientists, assuming that the anomalies surrounding explanation and confirmation will be settled by other specialists and, hence, allowing themselves speculations that proceed on that assumption.[24] Looking further, empiricists typically have an investment in technical and analytical tools that is directly threatened by recognition of the general failure of closure in empiricism. But the construction of axiomatized theories, the analysis of causality, and similar topics would still be of importance even if the pure epistemological stance of empiricism is let go, as we shall see. But there is the underlying point for many philosophers that the abandonment of the direct empirical anchor would threaten anarchy and relativism for scientific knowledge. This issue is probably crucial. The stakes in abandoning empiricism are so high from this point of view that they are equivalent to the intuition with which empiricism begins, the intuition that scientific examples of a certain kind are our best examples of genuine knowledge and must form the starting point of a satisfactory epistemology of science. A satisfactory postempiricism must preserve intuitions about these examples without threatening loss of objectivity.

To amplify this point slightly, suppose that it is conceded that background values and attitudes are coded into paradigms in order to provide scientists working in any area with a basic set of problems and kinds of solutions to these problems, but that these paradigms may undergo sudden and not entirely foreseeable shifts, as in Kuhn's description of scientific activity.[25] In our discussion of explanation, we noted that scientists are often constrained by intuitions about scientific plausibility, a view that would mesh nicely with Kuhn's account. Philosophical empiricism sees this view as threatening the objectivity of science, the conception that science progresses toward truth over time, and it even threatens the optimism for the future of science that is associated with empiricism. The slightest concession to internalized intuitive constraints seems to threaten relativism and even chaos. If the history of sociology of science suggests that scientists are governed by paradigms, the empiricist must see this as a lapse from normative ideals of disinterested experimentation and completely open logical criticism. Clearly the pressures toward remaining inside an empiricist's framework are quite strong for analytic philosophers who wish to retain a privileged conception of sound scientific advance, and the philosophy of science must respect these pressures.

A Dynamic Approach

In the history of science one encounters many cautious experimentalists who are suspicious of speculative science, and many bold theorists who are impatient with the repetition of clear experimental achievements. Both sides of the debate can score off recognized folly on the other side and develop a sound defense of their preferred position on the basis of key examples. Could science as a whole advance without bold speculation to point the way? Could science be trusted if it weren't for the cautious repetition and testing of results? Since the answer to both of these questions is clearly negative, there is something wrong with the attempt to reduce science either to bold theorizing or to cautious experiment. Without attempting simply to compromise between them, both activities will have to play a role in our account of science. If rationalism seems best to support theorizing, and empiricism best to support experimentation, they must both fall short of epistemological adequacy. For philosophers in these traditions, retention of their ideas must ultimately result in a considerable rewriting of scientific history. This seems too high a price to pay for philosophical theory.

Many attempts to avoid standard empiricist philosophies of science have not been able to avoid the conceptual framework of empiricism in accommodating the pressure from scientific history. Feyerabend, for example, has argued that there is no logical structure common to various scientific practices, so that the positivist attempt to find one would put blinders on scientists if they were to follow positivistic suggestions about method.[26] Feyerabend likes to consider as representative scientists figures like Galileo, who clearly violated any consistent methodological rules that might have been abstracted from the surrounding practice of his time, and who also violated in practice some standard empiricist notions of scientific rationality. In view of this situation, Feyerabend suggests a whiff of anarchism, and proposes that the concept of methodological rules be abandoned and that scientists and philosophers be encouraged to find and pursue as many alternative theories as possible to those now being developed. What seems here to be totally at odds with the standard view—a call to the greatest possible diversity and the scumbling of any clear line of demarcation between science and nonscience—actually shares many features with the empiricism it opposes. Each scientist is conceived as a solitary individual confronting amorphous data, free to theorize on his or her own. These scientists will still engage in standard forms of scientific argument and controversy, which Feyerabend expects to be usefully heightened by a more articulate diversity of outlook. Feyerabend's view that science (or society) should keep all possible traditions alive is politically naive.[27] No social theory is presented in his writings to make it at all clear how a multitude of traditions could appropriately be kept viable and yet in useful communication. Criteria and mechanisms for the protection of the weak are not elaborated. Feyerabend's advice is also historically naive. Some scientists, partly on competitive grounds, always seek alternative accounts to those currently reigning in any area of research. The real problem is to do this at an interesting level. Data frequently rule out all but one or two plausible alternative theories, and the advice to find more is ineffective without hints on how plausible alternatives are to be found. The scientists who don't pay attention to philosophy—and there are many—are hardly to be liberated by philosophical pronouncements of methodological anarchy. With modern big science, there may be economic silliness in proposing that every conceivable alternative tradition should be equally supported and encouraged. Perhaps just what is done is optimal for the rate of growth of science; whether true or not, this is just as unprovable as the advantages of anarchy. Anarchy or dadaism, while perhaps a healthy philosophical antidote to overly restrictive philosophies of science, doesn't seem to provide an account of science that brings us closer to answering our questions about the superiority of the insights of science into the empirical world, nor does it offer workable advice for improving or justifying scientific practice. Rather, anarchy or dadaism compromises empiricism and rationalism, but in the laziest possible fashion.

Kuhn's approach, so intensely disliked by empiricists, abandons any cumulative notion of the data to which theory is fitted, and makes data partially the result of observation that is determined by a background of paradigms. Revolutions occur in science, in which these background paradigms change; and after a revolution old data may be completely irrelevant to new paradigms, and thus there is no way to utilize them to evaluate the new theoretical outlook. Old data may simply be forgotten. For empiricism, there is no way to anticipate the content of data gathered by new instruments, and for Kuhn there is no way to anticipate the new theoretical outlook that may come into view when anomalies lead to an older outlook going under. In empiricism, even sophisticated versions, the data are perfectly consistent, determinate, and fixed at any point in time, at least after methodological laundering, and they determine the value of theories. In rationalism, including the variety offered by Kuhn, there is a fixed and determinate theoretical background that determines the value of data. There is give and take in neither. Popper allows one counterexample to prove a theory false, and Kuhn allows that sufficient anomalies will typically lead to adoption of a new paradigmatic background. What we do not find in either view is a constant accommodation of theory to experiment, or vice versa. Normal science as puzzle solving suggests wrongly that the pieces are given in advance, and need only to be correctly manipulated.

At any given time there will be experiments and theories or, as we shall put it more precisely, scientific facts comprising data domains and scientific theories. The significance of both cannot be given at the time when they are both first introduced and considered. What a fact is will depend on the development of theory to give it significance. What a theory is will depend on the development of scientific fact to give it significance. *Were the significance of facts given with their discovery, then empiricism would be correct, and theory would be dependent on fact. Were the significance of theories given with their formulations, then rationalism would be correct, and facts could only confirm the theories they were collected to test.* Both philosophies assume something in advance of further fact.

Empirical reality is too complex to be more than partially captured in any actual discourse, including human scientific discourse. All successful theories, as indeed all successful fictions, show us a way the world is without revealing *the* way that the world is, even partially.[28] They are thus false of total reality, while they may be true of some aspect of it. Our discourses describe an aspect of the world, and they describe the world most adequately at the rough level of our discourses, the level of human perception, not at the level of the very large or the very small with respect to that level. Rationalism, looking for comprehensive system, and empiricism, which supports typically the unified science hypothesis, share the conviction that empirical reality might in principle be fully described.[29]

A dynamic approach that transcends traditional empiricism and rationalism should not lapse into an easy relativism incompatible with our intuitions about the objectivity of science. It is not being said that all of an apparently incompatible set of theories can be true of reality; rather it is being said that all theories are (strictly) false of reality. Because of this, neither realism nor instrumentalism fits neatly into the view being developed. Insofar as theories show us a way the world is, a way we can expect to utilize in constructing new experiments, they are realistic. They work because they mirror reality, but their mirroring is always imperfect and can be improved. What can provide the principle of division of subject matter into controllable areas within which progress is a measurable achievement, even if the division is seen to violate philosophical desires for grand theory? Previous views have left the role of scientific experimentation out of account. A theory will be proposed to explain a range of data gathered by some fixed instrumental means. As these instruments are *refined*, so are the data, and theory will have to adjust to such refinement. Empiricism assumed that we can learn facts about nature, but instrumental means only produce a data text whose relationship to nature is problematic. A good fit with refined data constitutes success for a theory, and continued success is the clear indicator of progress. The ability to fit refined data constitutes success even where boundaries must be adjusted or auxiliary hypotheses added, provided that simpler rival theories do not materialize. Scientific progress will be measurable only for specific theories against such data domains on this view, and in these domains falsification cannot be sharp because the significance of apparently falsifying data cannot always be known when they are first taken from instruments. Over a period of time, the boundaries of a data domain may settle because of recognition of standard instrumentation for that domain. When new data are produced by new kinds of instruments, a new domain may be created for which quite different kinds of theories are needed. This fact will not erase the significance of the theories fitting established domains, thus allowing success of theory to have a permanent place in scientific history. These topics will be developed below.

The view proposed here is that fictions, including scientific theories, are not to be regarded as describing hypothetical worlds, but as describing aspects of a single, very complex reality. Scientific theories are fictions that in many cases anticipate aspects of reality to be observed later, but fictions that can be (counterfactually) regarded as true, just like various portraits and novels can be regarded as important insights into their subject matter. Although it may seem initially outrageous to assimilate novels and scientific theories as both revealing aspects of empirical reality, we can differentiate them. Theories, where they offer empirically valid constructions, are distinguished from other fictions in that the offered constructions satisfy criteria of experimental method and interpret important data text. To avoid a relapse into some form of vacuous rationalism, it is necessary to show how we can get outside the circle of theoretical ideas so that experimental results can be taken as a check upon and as causing revisions in, the theories and theoretical results that produced them. How is it that theories, if they control observation, do not reach down all the way into the data to make them useless except as confirmation of the theories that produced them?

If theory-laden observation cannot be avoided, we can still make it clear how observation can contradict theory. Unusual observations can be made against a background of everyday theoretical anticipations. Roentgen did not expect to see anything unusual when he developed the famous photographic plate, but he did. The very fact that language has a loose fit to the world allows such primitive observations of the unanticipated to be described, at least in graphic terms. At first such strange and unanticipated observations may seem difficult to analyze until enough theory is built up to enclose them.[30] In many cases, an observation is of no significance until a theoretical anticipation is developed that makes it significant. The bubble of rare gases at the end of experiments designed to filter out the constituents of air was seen as an artifact of an insufficiently completed experiment until the possibility of rare gases was conceived. This is why the title of a first discovery is generally awarded, not to the scientist first to confront a particular phenomenon, but to the scientist first to confront the phenomenon and *recognize* it (in an anticipatory fashion) for what it is in terms of the current theoretical framework.[31]

Now we can turn to everyday cases of experiments run to test a more or less explicitly formulated theory. First, a theory constrains observation by determining what to look for. When objects of different weight (properly: mass) were dropped in the seventeenth century to study their rate of fall, the theoretical background took only weight to be relevant to rate of fall, not, for example, color or shape. Objects have too many properties, properly an infinite number, for experiment to differentiate them and rule them out as relevant variables. A theory or theoretical concept makes a bold conjecture. It says what is relevant, and what to look for. Because of this, Galileo's theoretical background determined that he plot weight against rate of fall, and in this sense his observation was theory laden, but it did not determine that his plotting could come out in only one way. As experiments are repeated, they never produce exactly the same data in interesting cases. Thus the data aren't determined in every detail by prior theory. Only the kind of data to be sought is determined by theoretical expectation. Theory indicates what needs to be attended to in order for an experiment to come out right.[32] If it weren't for theory, we could never be even reasonably sure that we had isolated the right factors to observe, and that we were allowing them to run a significant course for our enlightenment.

Philosophers frequently confront theory and data, and either theory threatens the integrity of data because observation is thought to be theory dependent, or the data are taken to be fixed by observation independently of theory, so that theory must adjust to changing data and cannot evaluate, but merely explain, the data. If the data of science were typically subjective impressions or observations, then our reading of them would be suspect if our theory of the observations underwent a change. We might suspect that if we were to observe again, we would observe differently. The advantages of a scientific instrument are that it cannot change theories. Instruments embody theories, to be sure, or we wouldn't have any grasp of the significance of their operation. They can be taken to reveal a way that the world is in interaction with the world, just because their properties remain relatively fixed. Instruments create an invariant relationship between their operations and the world, at least when we abstract from the expertise involved in their correct use. When our theories change, we may conceive of the significance of the instrument and the world with which it is interacting differently, and the datum of an instrument reading may change in significance, but the datum can nonetheless stay the same, and will typically be expected to do so. An instrument reads 2 when exposed to some phenomenon. After a change in theory, it will continue to show the same reading, even though we may take the reading to be no longer important, or to tell us something other than what we had thought originally.

An interpretation of other sorts of text, such as Biblical materials or Shakespearean drama, deals with a fixed quantity of data represented by a finite amount of text to be interpreted. Clashing interpretations may be irresolvable, since each interpretation sees the text differently, but the text to be interpreted may be finite and fixed unless lost manuscripts are discovered. Such interpretation seems subjective to many, and interpretation itself seems based on insight or feeling, since there is no way to go beyond the given text to resolve disputes or to locate the reality that it describes. In science the situation is different, but not because the data necessarily measure reality, as traditional empiricism has claimed. The total text to be interpreted, the text of data, is not fixed before interpretation. The text of fact is constantly expanding, and can seemingly be endlessly expanded. Thus the clash of interpretations of fact, the clash of theory, is always potentially resolvable by expanding the factual text. What may seem objective by comparison to other textual interpretation because of this potential for the resolution of conflict does not require that language be circumvented in direct experience of nature. It is not part of the account offered here that one can step outside of scientific data to determine that they are actually true of the world. The primitive raw data of scientific experience have an unknown relationship to the world and are determined by the history of instrumentation.

The hypothetico-deductive account never produced a true dialectical interplay between theory and data. Because of its assumption that data were revelatory of nature, it always made data the test of theory and oversimplified the connection between theory and data. Our aim is to make data text as open to skepticism as theory, and to observe that advances in theory or instrumentation may render data useless or obsolete. Facts will not be traced to the level of individual experience, and they will prove to be as constructed as theory. This will allow us to avoid the simplified notions of explanation, falsifiability, or confirmation that have rested on simplified notions of scientific fact.

We have now made some progress toward discerning the elements required in a dynamic account of science. The interplay between theory and experiment is not simply an interplay between theory and subjective experience, but an interplay between theory and scientific fact, where a scientific fact is the product of an interaction between a scientist and the world as mediated by a scientific instrument in the typical contemporary case. That instruments will function similarly with scientists who hold quite different theoretical outlooks begins to point to a realm of scientific fact that is not dependent on subjective or intersubjective impressions, but much remains to be done. It will be argued that the history of instrumentation provides a unidirectional explanation of progress, in that later, more refined instruments are uniformly preferable to earlier instruments directed toward obtaining data in the same domain, and that this fact is essential to understanding the creation of what will be called data domains for scientific theory. Above these domains a complex dialetic of scientific fact and theory will hold sway, but the relevant facts will not be the impressions of individual scientists of the data in the domains, but publicly negotiated provisional fixed points for scientific argument about the significance of theory. What is left of empiricism is that these facts will play a coercive role in determining the development of theory.

· 2 ·

SOCIAL STRUCTURE IN SCIENCE

Social Norms in Science

One might suggest that the self-recognized best practice of scientists ought to be simply described, and this description offered as a characterization of scientific epistemology. Hanson has taken some interesting steps in this direction, but an immediate difficulty is apparent.[1] It must be assumed that not everything that even a great scientist does necessarily advances his or her best scientific work, no matter what the self-assessment of it is. An effort to eschew theory to the point where scientific practice is described but not evaluated, even the internal practice of scientific giants, points in the direction of a pile of examples of diverse scientific practice with no unifying theme. If, in the end, this is all that can be accomplished, then there is no point to the sociology of science.

Whatever the relationship between sociology and philosophy, the use of sociological material here is entirely subverted to the ultimate end of philosophical analysis. Perhaps this will allay the xenophobia of philosophers who are too quickly dismissive or contemptuous of other disciplines. Sociological description is requisite for the social structure of science to become evident enough to allow for the development of a dialectical scientific epistemology. Ultimately, however, there is a failure of such description that is not unlike the failure of closure encountered in traditional philosophical epistemologies. Just as the traditions of empiricism and rationalism cannot both explain important features of science—the crucial roles of *both* theory and data, for example—so the sociological tradition will fail to explain some important features of science. In seeing science as an institution, or as a culture, the revelations of sociological analysis must deal inevitably with features of science that make it similar to other social institutions. Other small human groups organized around some mutual concern that is not scientific in the intuitive sense may ape the social structure of scientific research groups. The problem for the sociology of science is to find a true *differentia* of scientific practice within the academic disciplines.

We can begin with the view that science is a social institution among other social institutions, and then apply various sociological techniques to study science as an institution. Scientists can be counted, their citations of one another noted, and in this way a certain amount of factual material becomes available to the student of the social structure of science. The pattern of scientific social structure obtained in this way is very hard to differentiate from other intellectual social structures associated with nonscientific academic life. Literary critics make discoveries, cite one another, abandon various theories under the pressure of new evidence, and in many ways comport themselves in terms of social structure and private epistemology like scientists, except, of course, that they don't on average have as much disposable income, and they are not generally thought to succeed in locating objective knowledge. The former is an accidental fact that can hardly explain the latter.

This external structural approach to science can be sharpened by looking for the institutional norms of scientific behavior, violation of which produces sanctions by other scientists. An early and widely cited attempt to do this was made by Merton, who found a set of norms that he took to be characteristic of science from the seventeenth century to the present.[2] Merton has been attacked on the basis that the norms of science have noticeably changed over this period of time, but another issue seems much more important.[3] Although Merton's norms are grouped differently in different texts, making quotation difficult, a fairly differentiated list can be cited as follows:[4]

Faith in rationality
Universalism (All scientists have equal claims to the discovery and possession of rational knowledge.)
Individualism (Science is antiauthoritarian.)
Community (Credit for discovery is given to individuals, and secrecy about results is immoral.)
Disinterestedness (Self-interest is achieved through community recognition. Financial gain through discovery is wrong.)
Impartiality (The scientist is interested in pure knowledge, not in applications.)
Suspension of judgment (Evidence is the only arbiter of truth.)
Absence of bias
Group loyalty (Production of new knowledge by scientists is supported over guidance of research by economic interests.)
Freedom (Control of scientific investigation is regulated only by epistemological considerations.)

The norms are somewhat condensed and overlap too much in any such summary, but their general import is clear. Certainly these norms have some relationship to science, and yet it is manifest that they are frequently violated in fact, and that violation need not lead to the imposition of sanctions. It would, however, be a mistake to dismiss these norms as simply providing an overly idealized and mistaken picture of science. These norms portray the picture of science that many scientists prefer to see as the public image of their profession, and that many scientists believe they follow when *they* are doing science. But these attitudes may betray a lack of self-consciousness on the part of the scientists evincing them, although this is perhaps difficult to see at first. This set of norms can be thought of as standing to science as its ideology, and the attitude of many scientists to this ideology can be thought of as a (necessary) form of false consciousness that aids concentration on scientific work.

To see what is meant by this, we can imagine that some medieval theologian is working on an interpretation of, let us say, a book of the Bible. In doing this, he may (privately) want to show that his knowledge of the Bible and his insight into its meaning is greater than that of any of his rivals. His quest for status is not undertaken through physical violence against his rivals, nor can he be successful just by claiming greater insight. There is a physical and largely determinate text to be dealt with, and a variety of background constraints of an intellectual and theological nature to be satisfied. Success depends on satisfying these constraints in a manner that a large group, if not a majority, of interested scholars and church authorities will recognize as superior to attempts at interpretation by the theologian's rivals.

The case of the scientist with respect to rivals seems not all that dissimilar. A scientist will pick a problem that seems soluble to him or her, given background, available equipment, and so on, as well as one whose solution is estimated to be associated with the desired recognition and status. So much goes without saying, and the scientist may not consciously recognize this fact, simply because everyone behaves in a similar way. As in the case of the theologian, success depends on background constraints not within the control of the individual scientist. There are recognized experimental standards, a background set of plausible theoretical possibilities, and there are standards of scientific exposition and debate to which the Mertonian norms are relevant. Scientific status is sought by a procedure whose social interaction is governed by the Mertonian norms. A scientist may be vain and arrogant, but his successful public *performance* depends on treating all other views only in terms of their experimental support and theoretical consequences. Public science and public theology are then played out against a set of recognized professional standards, standards that are not to be compromised by the personal goals of the participants. The Mertonian norms are, so to speak, the rules of the game.

The Mertonian norms have in recent years been confronted by the recognition by some scientists that human motives play an essential role in scientific research. Watson's *The Double Helix* is an outstanding example of exposé literature, and was perhaps the first book to explicitly acknowledge the role of personal goals in motivating scientific work.[5] In this autobiographical account by one of the discoverers of DNA structure, jealousy that a distinguished scientist might achieve the result first, spying on that scientist's work through an intermediary, and outright secrecy about personal results all play a role. But these factors do not contradict the Mertonian norms, because the personal factors played a role in motivating work on the discovery, but not in the final public communication of the discovery and the subsequent public scientific discussion. Science has been enormously successful in preserving a distinction between the public, scientific agent and the private individual. Scientific results are communicated and discussed quite independently of the religious, moral, political, or philosophical opinions of the participants, although these opinions will influence choice of research, style of argument, and even initial judgment of work. While some scientists might have been embarrassed by Watson's lack of taste, his work was unaffected by that fact, and could be used by any other scientist.

The fact that the Mertonian norms rule public science is quite consistent with greed and passion, or impartiality and fair play, at the personal level. Ravetz has suggested that the recruitment of scientists from wider segments of society is having an influence on the type of person who is becoming a scientist.[6] Although motivational factors in the performance of the highest-quality scientific work of which one is capable are subtle and complex, a sheer shift in personality type is compatible with continuous quality in scientific work, provided that the public norms of conduct can be preserved in scientific controversy.

In many areas of active research in science, pairs of scientists can be found who disagree on some specific proposition. Typically, the scientists who disagree will make quite different simplifying assumptions in order to construct the models from which they draw conflicting conclusions. Each scientist in such a dispute may believe the other scientist to have made mistaken or unilluminating simplifying assumptions, but as all simplifying assumptions are recognized as such, and as the full range of consequences may not yet have been drawn from these assumptions, it is not necessarily part of the scientific dispute to quarrel about these assumptions. Each scientist may be primarily concerned to show the rigor and thoroughness with which his assumptions have been pursued, while recognizing an arbitrary element in the assumptions themselves. A detailed analysis of one such example by Martin is quite illuminating in this connection.[7] In his example, he discusses the presuppositions found in two apparently conflicting scientific reports on the potential hazards of certain emissions from supersonic transport planes. Martins's analysis, communicated to the authors, was that their assumptions about what was to be proved had led them to select different models for emissions, and to quote other sources in a manner somewhat influenced by their intuitive judgments. Both authors denied this vigorously, claiming that their choice of model had been based on perfectly reasonable simplifications, and that they had rigorously drawn the implications implicit in reasonable starting points.

Mertonian rules thus seem to bear an analogy to proper courtroom procedures between adversaries. They are not necessarily to be found in the individual pursuit of scientific knowledge, but they do constrain the cultural products of science, research papers and books, for example, and they also constrain the public debate at the heart of scientific progress. Mertonian public norms revolve around full disclosure of experimental results, or at least around the idea that no false statements about experimental results will be publicly disclosed. Scientists are tied into a network of other scientists, and the work of other scientists must be trustworthy if the system is to continue to develop at a satisfactory rate. Mutual trust would be subverted by cheating. A scientist could manufacture results, or not reveal results that were incompatible with his or her favorite results, and this has happened. But on the whole, scientific morale has remained high, and cheating (even if it is increasing) has been rare enough to provoke scandal and notoriety where it has been discovered. What is the mechanism that brings this about? The mere pledge to uphold Mertonian ideas or the assertion that these ideals are to be observed in public discussion is not explanatory. Pledges that are at odds with the exigencies surrounding personal gain will inevitably give way to prudential calculations. Workers in a factory may make various work pledges quite at odds with private intention, and break the pledges at the first reasonable opportunity. Workers may desire the most pay for the least possible effort, the reverse of the employer's objectives.[8] On the other hand, scientists consistently strive for the best work they are capable of, and we need to find the differentia between the professions and the scientists as well as the differentia between workers and scientists.

Two factors seem of crucial importance in explaining scientific motivation to do the high-quality work that is consonant with institutional goals. One is that a scientist works at a single problem for a pretty long period of time, typically, and can be expected to make only a few significant discoveries in a lifetime of work. The relationship between the scientist and his work is thus found on a time scale where a scientist can expect to take personal pride in a good piece of work. The discovery or result will be his or hers, given the reward system of science. Scientific work thus shares a relationship with artistic work, or craft work, in which the signature of the producer is regarded as part of the work and may function as a sign of worth and quality recognized by others. If quality work is the goal of science, the reward system for the individual is consistent with the institutional goal, and puts pressure on the individual to produce work that can be defended in the public arena as a contribution to science for which he or she should receive recognition. The other important factor is that science embodies a social structure in which each individual can imagine a constant chance for personal recognition. From one point of view, science possesses very little hierarchy, It is fragmented into research areas within which there are recognized stars (of perhaps approximately equal magnitude) and the rest of the researchers. A Nobel Prize winner may be an important physicist, but the same prize winner may also be irrelevant to a research area in which there have been no Nobel Prize winners. Workers in this area need not see the prize winner as standing above them in a hierarchy. Each scientist may thus imagine that one good piece of work is all that it would take to achieve scientific eminence, at least among those working on some well-defined research topic. By comparison, the factory worker is typically resigned to remaining in his or her position, or one like it, since the pyramid of company offices leading to president is a constant and visible reminder of the difficulties of rising through the hierarchy. In the professions, the typical professional is practicing the profession as a source of superior income, and no system of recognition (except for local social hierarchies) exists that would motivate the search for excellence in practice. Excellence under these circumstances must be pursued on personal grounds, or because it is somehow expected to attract superior clients in the long run of one's career.

The social structure isolated here does not begin to explain the objectivity of scientific knowlege. It is designed merely to illumniate why cheating is so scarce in science, that is, why individual and institutional goals are so remarkably consonant in practice. Merton's norms are what would be expected in a system of such consonance on individual recognition. We can say that in the past, science as pursued primarily by individuals maintained a reward system in which the possibility of recognition for everyone who did good work was complemented by individual attempts to do good work. The institution and the individual were in harmony. Structural changes seem much more threatening to this system than any suspected lessening in the personal qualities of scientists. The cheating scientist of the past had to restrict himself or herself to minor league work, or risk a substantial chance that dishonesty would be discovered in repetition of the works, when the original results could not be defended, since the thread of responsibility was securely anchored at one end in a particular person.

The move to big science threatens the stability of this system. Within a scientific team cheating may be encouraged in lower positions because of a political pressure to please one's superiors. In the team situation, the lower positions will often be filled by persons knowing that the recognition for success will go primarily to team leaders. They thus lose the pure motivation to do good work, and are subject to political suasion, unless they belong to a team whose leadership creates the atmosphere for top-to-bottom quality, and whose recruitment of talents and personalities is lucky. In those research areas where large funding is requisite, those scientists or teams of scientists who are not able to secure funding essentially lose their chance for recognition. Recent years have seen squabbles and studies of the way in which scientific grant monies are administered. There is not enough money to fund everyone who would like to do expensive research, and those who are not funded have to shift research interests in many cases. Attention under these circumstances will be shifted to research projects that have a good chance of being funded, rather than those that might be chosen on purely scientific grounds. It is possible, of course, that these motivations could coincide, but frequently they will not. Whitley reports that a physics research laboratory receiving government funding may spend up to half of its time replicating experiments in order to produce research reports, such reports being the government's criterion of scientific productivity.[9] In this way, time and labor can be deflected from pure research. For those who are losers in the funding game, morale may be difficult to sustain. Big science, necessary for economic reasons in many areas of research, threatens the personal award system that seems to have played an essential role in preserving the consonance between personal motivation and scientific goals that has characterized the explosion of scientific knowledge for over two centuries.

It may seem that this brief look at social norms in science has not advanced our topic at all. If science is seen as a public struggle for recognition played out according to certain internal social norms, is it not true that this is true of all academic disciplines? Let us return to our notion of science as containing two levels of texts. In the humanities, the important texts are set by tradition, they are small in number, and the scholar is always confronted with several existing and presumably plausible prior interpretations. Even when a new interpretation is added to this array, choice among them depends on feel and insight. One cannot force recognition of the validity of one's interpretation. It is possible, of course, to try to write new texts, but the vagaries of fame tend to preclude this as a reasonable route to success, and simply attempting to fix up flawed but existent texts is not regarded as original, even where it is legitimate. In science, opportunities beckon. One can write new text, interpret new text, or pursue a variety of other activities such as emending or otherwise improving existent text. The existence of these opportunities would not only explain the historical high quality of scientific work in all fields of science, but also offer a basis for a distinction between scientific work and work in other academic disciplines.

Cognitive Norms in Science

A different sociological approach to science is to regard scientists or groups of scientists as constituting a special society or community with its own associated culture. Science is not then regarded as an institution within a given society, but as constituting a society in itself. Instead of comparing the institution of science to other social institutions, one could then attempt to describe the internal culture of science and scientific subgroups. Kuhn's well-known book, *The Structure of Scientific Revolutions*, approaches science in this way, although for Kuhn science turns out to be less a total society with an associated culture than an assemblage of fairly small research societies or groups each of which is unified in direction and values by background paradigms.[10] Like members of any culture, the members of a research group cannot fully articulate the sources of their unity of outlook, since the consensus is partly the result of sharing a research activity that is not itself fully and consciously spelled out in all of its details. If this view that the sources of consensus can't be articulated is correct, incidentally, it spells defeat for any program attempting a neutral description of scientific practice. The fragmentation of science that results from this view is inimical to philosophical generalization, since the practices of each research group will make sense only against its particular paradigmatic background. Scientists are constrained in perception by this background according to Kuhn's account, and perhaps this is why Kuhn's approach has not been widely admired by philosophers.[11] They tend to see scientific activity as no longer fully free and rational if it is governed by paradigms, since paradigmatic backgrounds may be taken to function as an inarticulate and hence irrational authority. The moment a paradigm could be spelled out, it could (and presumably would) be open to rational discussion and consideration of alternatives, and would have lost its constraining force. Paradigms, if they could be articulated, would be acceptable to philosophers, but they could no longer play the role in scientific research that Kuhn spells out. At the same time, many scientists have welcomed Kuhn's account, and they seem unmoved by the philosophically ominous inarticulate properties of a paradigm. Their willingness to accept Kuhn's account may be partly because Kuhn's account makes scientists a breed apart, and scientific expertise something that can only be achieved by a long process of training that resembles a process of acculturation. Pure scientists, especially, like to view their work as autonomous with respect to other forms of social activity, and Kuhn's account allows them this luxury.

The scientific norms established by paradigms are basically cognitive norms in the sense that they constrain scientists in their cognitive activity toward a consensus on the significance of formulae, instruments, and experimental data. In practice, Kuhn's exact account of paradigms has wavered somewhat. Originally, Kuhn took paradigms to be the entire background set of values and attitudes of a scientific research group, as well as the particular formulae and exemplars (standard ways of connecting theory and formulae to experimental arrangements) characteristic of the research group. Lately, the sense of the term *paradigm* has been narrowed to that of *exemplar*, but the general conception of consensus in unifying a research group remains the same. For Kuhn, paradigms constrain the members of a research group to act similarly in approaching nature, and to expect their interaction with nature to proceed along certain anticipated lines. This allows a division of labor to occur in which scientists can cooperatively pool their efforts because they needn't argue about fundamentals, hence allowing progress to occur. Greater consensus over paradigms, as in physics, allows homogeneous research groups to participate in an accelerated attack on common problems.

During periods of normal science, scientific activity proceeds by solving new problems in the manner sanctioned by the paradigm, and in developing experiments to extend the paradigm's range of application. This has the interesting twist that the power of the paradigm to constrain thought means that a result not fully expected within the constraints of the paradigm is at first taken as a signal to the scientist that the experiment involved was poorly designed, not that the paradigm might be wrong. This observation jeopardizes the simple logic of falsification that philosophers find so congenial, and means that the reaction of good scientists to an experiment can't be predicted by mere logic. The idea that an outsider point of view could determine how scientists should react if they are fully rational is irrelevant to the actual practice of science. What scientists see as plausible and relevant is determined by a space created by a paradigm, not by a space created by logic. This account is therefore compatible with what was said in the last chapter about logical space in connection with scientific explanation. To understand a research group, some knowledge of the content of its paradigm or paradigms is essential in the Kuhnian reading of science as scientific culture.

Reactions to Kuhn have been sharply divided between complete rejection sometimes expressed in various reasonable objections to the sketchy account of research groups and paradigms offered in his book, and acceptance followed by an attempt to spell out the relevant paradigms for various research groups, or even for whole disciplines. There is no doubt both that Kuhn's account was the first really insightful description of everyday scientific activity and that it contains serious flaws. The Kuhnian account of research groups suffers from the general drawback of functionalist descriptions of social groups, which tend to make such groups seem both more uniform internally and more distinct externally than they actually are. There is in fact both violent internal disagreement within the research groups met in science and hot argument between research groups ruled by different paradigms. The former is bloodier than paradigmatic unity would suggest, and the latter ought not to take place with the vehemence it does if adherents to different paradigms should find each other's attitudes and work incomprehensible and even nonscientific. Because research groups are too internally uniform on Kuhn's description, he is forced to hold that when anomalies pile up sufficiently against a paradigm, only a revolutionary escape to a new paradigm is possible, and once this is accomplished, or once new people have populated a new research group, rational communication between the old and new research groups would be impossible. Kuhn's account, based on total agreement inside the group, leads inevitably to this end. It is clear that his original account overemphasized the uniformity of opinion within a research group, and that a more evolutionary account of progress is required in which scientists within research groups sharing the same paradigmatic exemplars may have markedly different outlooks, and in which talk between research groups is possible, including fruitful discussion between scientists who are utterly at odds about the assumptions and models involved in their work. An improvement on Kuhn requires seeing that the individual scientist usually doesn't have grand goals except in philosophical moments, but rather he or she is working on some quite specific problem to which a solution can be hoped for in the near future. To solve it, the scientist may try a wide variety of approaches in a relatively short period of time, and change his or her theoretical outlook repeatedly, something that is not quite compatible with Kuhn's original account.

The main stream of scientific history, as well as the large revolutionary shifts in scientific opinion that occur occasionally and affect many scientists, has to be reconstructed from a detailed history in which people are not quite sure where they are going, or how they are going to get there, or what the significance of their work really is. That a more evolutionary account seems to fit scientific practice better is shown in various studies of scientific specialties, of which one by Crane is extremely interesting.[12] Crane looked at theoretical high-energy physics during the fifteen-year period 1960–1975. Physics, of course, should exhibit Kuhn's account most clearly, since he developed his account partly to explain the apparently rapid progress in physics in comparison to other scientific disciplines. Theoreticians during the period in question sought to explain what they called weak interactions, strong interactions, or both. Crane found more than twenty different lines of inquiry into these phenomena to have attracted theorists during this fifteen-year period. She also found a quite complicated pattern of influence between these lines of inquiry, with scientists having shifted allegiances repeatedly, partly because of experimental results, and apparently partly because of sheer fashion and intellectual boredom. During this period there were several unexpected experimental findings, but only one real anomaly (charge and parity violation). This tends to cast doubt on the idea that theoretical progress comes through periods of revolution, and that revolutions are a response to a large stockpiling of anomalies. Further, Crane found that the entire area of theoretical high-energy physics accepted certain symbolic generalizations related to relativistic quantum mechanics, and also accepted certain general values (testability and elegance) for evaluating theories. Individual scientists in the area accepted various metaphysical models quite independently of their line of inquiry, but exemplars (in Kuhn's sense) were held in common by those involved in a given line of inquiry.[13] Crane's research suggests a much richer pattern of grouping than that proposed by Kuhn.

Kuhn's method of description emphasizing paradigms approaches a phenomenon we noted earlier from another direction. The thinking of scientists is highly constrained by background scientific assumptions, not all of which are totally conscious, and this may seem irrational from a philosophical point of view, unless it is realized that these assumptions are made at a level of particularity that has been worked out in close interaction with the phenomena being studied. Scientists participate in some diffuse cognitive attitudes, such as a faith in the fit between mathematics and the world, a belief that fundamental processes can be understood by the human mind, and so forth. Also, because of the way in which the sciences are divided in the universities, the journal systems, grant-giving agencies, and so forth, scientists will see themselves as physicists, organic chemists, and so on, and will share some disciplinewide attitudes. All of these background factors will play some role in scientific thinking. But at the daily level of work, the important factors will be that so-and-so has obtained such and such results with this method, that so-and-so has obtained what seems a suspiciously accurate result using a device that normally gives results difficult to replicate, and so on. That is, day-to-day scientific thinking will be most constrained cognitively by a knowledge of what other scientists are doing, what processes have brought what significant results, and so forth. The notion of a paradigm is a convenient way to capture this, but the associated picture of science as fragmented into internally consistent research groups ruled by paradigms who find one another incomprehensible and who occasionally destruct when too many anomalies pile up, to be replaced by new groups, is simply too crude a sociological picture of science to serve as anything other than a first approximation. As a first revision, we need to note that although paradigms may constrain the direction of research quite decidedly, research groups will often contain mixtures of individuals who disagree on various matters (how to go on) even though they do agree on the values ensured by the shared paradigm. For example, for one scientist a new and more careful experiment testing an old idea may seem preferable to the experiment testing a new idea favored by another, even though both can agree that with sufficient time and resources, both steps would be desirable.

Further, and this is crucial, a mixture of divergent thinkers may be quite essential for the stimulation leading to creative advance.

Kuhn's account overemphasizes the power of pardigms to coerce thought. He gives examples of how hard anomalies are to notice, and compares this process to perceptual studies in which expectation makes it difficult for a subject to notice certain things.[14] If theory determines observation, and if paradigms determine thought, how can anomalies be noticed? The answer to this puzzle lies with a human ability to grasp several paradigms more or less at the same time. Just as an anthropologist may be native to several cultures, and have the ability to evaluate some event from several different cultural perspectives, so a scientist may be native to several paradigms. Scientists move from research group to research group. Two scientists from different prior groups joining the same research group will not internalize exactly the same paradigm. They may construct new paradigms out of a sensible fusion of elements of the old and the new, so that paradigms need not spring full-blown into view after revolutionary periods, and scientists may come to realize that one paradigm is more globally satisfactory than another, even though the view from the two paradigms is quite distinct.[15] Kuhn, by making paradigms totally distinct from one another, makes it seem unlikely that one could lose its grip on a person and be replaced by another. Kuhn's scientists are as dull as Cartesian robots, consistent machines lacking any capacity or instinct for playing with a variety of viewpoints. This seems false to human psychology, and to the fact that scientists successfully migrate from one research group to another and require only a short period of time to readjust. The correct insight is that paradigms are specific to problem areas and are not fully articulate. This allows a small group to communicate fully through language that is partly symbolic, since its meaning depends on a context not shared by outsiders to the practice. This full communication implies a small group, common experiences, common understanding, and perhaps a special language coding these shared features. This shared full communication seems a requisite for scientific advance at the microlevel. Any attempt to spell out the paradigm, say, at the level of a *Scientific American* account of a discovery, loses this symbolic dimension of group language, and makes the thought of the group seem simpler and more obvious than it really is, since the shared subtleties of understanding, what is permitted and what is forbidden by the techniques, the data, and so on, are either stripped away or lose plausibility in gargantuan formulations. The same process occurs in the preparation of the research report, where full disclosure of background assumptions would swamp the vital information in a sea of irregular assumptions well enough known to those for whom the report is written.

Much of the discussion surrounding Kuhn's views has been related to the fact that the level at which they are intended to be applied remains obscure in his work, and no precise formulation seems to fit all the relevant epochs of the history of science. Copernicus, Darwin, Freud, and Einstein are all associated with scientific revolutions that originally occurred within small research groups, but had far-reaching consequences for views held in everyday life. Many people found these views difficult to accept, and many people never accepted them. Yet they do fully deserve the title *revolutionary*. It is not true, however, that no older scientists could accept these new paradigms, and that they triumphed only when they were adopted by younger scientists, as Kuhn suggests in his book.[16] All of these revolutionary ideas were accepted by at least some scientists in every age bracket, and even though older scientists were dying off at a higher rate than younger scientists during this period, there is no reason to suspect that ideas kill older scientists, or that the death of older scientists facilitates the spread of ideas.[17] Age cannot be even the major explanatory factor for acceptance or resistance. To be sure, older scientists are likely to have intellectual capital invested in older ideas, but on many grounds they might choose to accept a new set of ideas, even on the grounds offered by reason and argument. For younger scientists, the seemingly natural view that they are inclined to leap toward new ideas encounters problems. A young scientist seeking to establish a reputation might decide either to work in the area of a new paradigm in order to challenge received opinions, or to make some piece of work more rigorous than ever before by carefully redoing the past. On the whole, the young are likely to differ from the old, if only because of the imprecision of education as well as the fact that education is set against a different social background over time. A modern scientist might consciously choose to pursue some work of ecological significance without comment, but such a choice would not have been so natural to contemporaries fifty years ago. It is hardly credible to assert that new Ph.D.'s do not have minds that are as well stocked as older minds, and hence are likely to be attracted to more speculative ideas because they don't perceive the lack of solid scientific support for them. Clearly, a young scientist may choose to pursue a wide range of strategies in developing a scientific career. If age does not play as important a role in connection with revolutionary new paradigms, as Kuhn has suggested, it becomes of questionable significance in a modern setting, where a scientific revolution has significance primarily for valuations within some scientific specialty, and hardly confronts major world views. Here a successful new paradigm may reorganize thinking or open new avenues of research in a manner accessible to those working in closely related areas, and rapid adaptive switches and conversions may be expected along with internalization of the new paradigmatic outlook. Much of the apparent excitement of Kuhn's views seems to result from an unnoticed transfer of the significance of the few huge revolutions in scientific thought that affected human world views in an earlier age of science to revolutions now occurring in connection with research groups within ordinary scientific advance. These are the locus for Kuhn's considered views about scientific progress that do not entail wild relativism or incomprehensibility theses, mostly because paradigmatic views are located within the confines of disciplines that have research groups oriented to recognized objects of inquiry, settled styles of experimental and theoretical attack, and a variety of associated stabilizing values. Relativism and incomprehensibility will be found in research on extrasensory perception, UFOs, and wherever a disciplinary anchor is not available and people may approach a given problem from any direction whatsoever.

One thing Kuhn has in common with his philosophical critics is a reliance on the idea that the history of science can be written as a history of scientific ideas and the way in which they have changed over time. The succession of scientific ideas must be related to the succession of scientific instruments, and without such an underpinning in data the notion of shared paradigms and exemplars cannot be fully fleshed out. What seems possible and impossible to scientists, especially where this is intuitive and only partly articulated, is frequently linked both to actual experiments and to the agreed results of thought experiments in some research area, and these constraints will not appear explicitly in the theories or papers of a research group, nor need they be agreed on by all members of a research group. Quantum theory is a good example of a physical theory that has developed to fit a space provided by experimental results that at first seemed odd, and then came to be the only results expected.[18] Here accumulating experimental results seemed to have gradually forced a quantum consensus outlook onto scientists who felt that they could interpret the experimental devices classically, and who gradually came to share similar attitudes about interpretation of the data obtained from these devices. This is quite the opposite of a consensus forming that was then elaborated in experimental testing. An important feature of scientific instruments and experimental techniques is that they reduce the variety and complexity of actual situations to a few manageable results, most of which can be numerically recorded. If we grant that observation may be influenced by background theory and background theoretical expectations, the fact that observations are transferred through instrumentation serves to detach the value of observations from the hopes and expectations of the scientist. Rather than making simple interpretations of complex objects, scientists tend to make complex interpretations of relatively simple objects. Instead of reacting to an entire process, a scientist reacts to some numbers or some feature of a process that he will accept in common with other scientists who may have quite different interpretations of the same process. The atheist and the theist react to their entire experience of the universe with conflicting claims, but two scientists may react to a single number representing some experimental datum by tracing quite different consequences from it in the context of two quite diverse and complicated theories. Instruments and techniques assure scientists that they are talking about the same thing, that is, some scientific fact, when they disagree. On the other hand, it is questionable whether the atheist and the theist inhabit the same universe. Positivists sought for certainty in scientific observation, but they overlooked the possibility of grounding community of understanding in science on instruments and techniques. The progress of science is really the progress of instruments and techniques. Better instruments and techniques tell us more about the same universe, and hence they frequently force changes in theoretical outlook. In this sense, there is positive progress in science even though theoretical interpretations may undergo wide flip-flops, what is taken for granted at one time being questioned for a while and then taken for granted again.[19] At times, new instruments and techniques may reveal unanticipated constituents or aspects of the universe. This is why progress isn't cumulative, and data may become worthless. At times, new interpretations may shift the significance of data, making crucial and important certain data that were previously considered of marginal significance, but this will not in itself change the values of agreed data. The fact that perception is determined partly by expectation is combated in science through instrument and technique, which tend to establish data that can stand independently of theoretical outlook, even though the data may not be neutral on which theory can most likely be projected onto new data in an area of research.

Kuhn's view, like that suggested in the last chapter, entails that there is no sure test for progress at the microlevel of scientific practice. Paradigms replace one another, but one cannot say whether a given replacement is a mistake or not, or whether it gives greater insight into the workings of nature. Later on, the path to the present will be clearer, and what is progress can be measured as significant steps toward the present. In response to critics who charge that this account makes scientific knowledge relative or subjective, Kuhn has suggested that progress just is what happens in scientific history, so that criticisms based on his failure to provide a criterion of progress based on gradual approximation to the truth fail because the critics are also unable to specify what approximation to the truth can mean, and are hence unable to provide a test for progress at the microlevel of scientific activity.[20] But science can't really be taken to progress as a result of a series of revolutions in thought, and both Kuhn's view and that of his critics have overlooked the importance of setting ideas into a context of scientific experimentation. That scientific history is not punctuated by revolutions in Kuhn's sense is evident from the fact that older theories have survived revolutions to take a place in science along with newer theory. A good example of this is that Newtonian physics is still a good physical theory and is studied by students of physics even though Einsteinian relativity theory has replaced Newton's theory in many physical applications as a deeper insight into nature. Similarly, classical physics has survived alongside quantum physics, and in many cases is used along with quantum theory in the explanation of experiments, although, once again, quantum physics is regarded as providing deeper insights into nature. Both of these pairs of theories are logically incompatible even though one may loosely be regarded as an approximation to the other under certain circumstances. It is sometimes argued that although the approximated theory is false, it is used for pragmatic reasons to give suitable answers when great precision is not required. In contrast to Newtonian and classical physics, however, phlogiston theories have disappeared. The difference once again requires a reference to the missing factor of instruments and techniques. As long as naked-eye observations and simple techniques and instruments are employed on human-scale objects at low differential velocities, in other words, as long as the data are restricted to the only kind of data that could be gathered by scientists given their instrumentarium until the twentieth century, classical physics is a vast, conceptually simple, interlocking, and consistent system expressed in a language that is perfectly precise and accurate with respect to the suggested data.[21] The phlogiston and impetus theories were set against a quite modest and imprecise data basis by comparison. It is no wonder that they were less robust in the face of new data and competing theoretical explanations.

Scientific Disciplines

The pure scientific disciplines seem easy to list in terms of their relationship to the modern university curriculum. Physics, chemistry, and biology (possibly geology) are clearly scientific disciplines, and both psychology and sociology claim scientific status as social sciences. Branching from chemistry and biology are the partly applied disciplines of biomedicine and agriculture. Mathematics may or may not be regarded as a science, but applied versions occur as disciplines with putative scientific status in engineering and economics. History and political science may also lay claims, at least in certain forms, to scientific status. This basic structure is, of course, the structure of university science, and the identity of scientific disciplines is tied up with the extensive development of modern science in the nineteenth century as part of the modern university. At present, chemistry, engineering, and biology departments are usually divided (at least in the larger universities) into departments constituting such subdisciplines as polymer science, electrical engineering, and botany. The inevitable ossification of university structure has made it difficult for some new disciplines seeking scientific status, for example, computer science, to find a clear place in the university structure. Sophisticated analyses of citations and key words in journal articles tend to support this well-known informal structure of disciplines, and that is not surprising, since both scientists and their journal editors tend to view scientific work as grouped into these traditional disciplines and associated emergent subdisciplines.

It is clear that one could begin research on disciplines with scientific papers and analyze the references to past work contained in them. The details are complex, because the pitfalls are nearly obvious to inspection. Explicit reference to sufficiently well known work is not necessary among scientists, and it is not clear when explicit references indicate a genuine intellectual debt or conceptual linkage, rather than attempts to take a wide range of literature into nominal account. Despite problems related to these and other observations, different studies have found the same gross structures in the literature. If one starts with an intuitively related set of papers, citations in the papers tend to close in on themselves so that papers can be grouped in clusters whose authors constitute research groups in the sense that we have been mentioning. These clusters can in turn be subdivided by strengthening the citation criteria required to code influence. By studying the literature from a selected list of journals in chemistry, physics, and biomedicine, for example, investigators have found that physics is largely separate from chemistry and biomedicine, but biomedicine and chemistry have an important overlap in which, for example, literature on the structure of hemoglobin is not clearly chemical research or biomedical research in the eyes of its participants.[22] Biomedicine, in fact, contains a huge number of clusters that are harder to separate, since biomedicine contains many review and method articles that are cited by clusters not otherwise citing the same literature, and these must be ignored to achieve cluster separation. These studies also show that chemistry lies between physics and biomedicine in that the links between physics and biomedicine are not as strong as those between physics and chemistry and between chemistry and biomedicine. Citation research confirms the standardly perceived structure of scientific disciplines, and also confirms the Kuhnian insight that scientists (and their documents) can be clustered into small groups with high cognitive cohesion within the groups, the members of which are attacking some fairly specific scientific problems, although citation research can't by itself confirm the idea that Kuhnian paradigms cause group cohesion.

What is somewhat more surprising is that intuitive differences between scientific disciplines often thought to exist do not appear clearly in the data when these disciplines are studied using citation research. For example, many people hold the opinion that physics is a paradigm science, but that sociology stands somewhere between physics and philosophy or between physics and literary criticism. One aspect of this belief is the view that physics has progressed at a much greater rate than sociology has in the last century. A measure of such a belief would be the age of citations. One would expect citations in physics to be to papers that are not very old, so that recent work in physics is more likely to be cited than older work, but statistics measuring this phenomenon show that all of the sciences have a similar reference pattern in terms of the age of citations, and there is a sharp break only between all of the putative sciences and the humanities in terms of reference patterns, the humanities averaging much older reference dates than the sciences do.[23] Within disciplines, subdisciplines and research groups can vary considerably in citation patterns. Certain areas of experimental psychology will tend to cite only very recent work, and other areas, such as clinical psychology, will cite at least some older work. This pattern of differing citation age in subdisciplines is true for all of the sciences, although disciplinewide averages vary little. The result is that the scientific disciplines show a remarkably similar citation pattern in statistical analysis. The differences between the sciences, or even average differences in some sense, that are intuitively thought to mark cognitive distinctions in paradigm structure between disciplines seem (at least so far) to elude statistical technique.

The social structure of scientific disciplines we have analyzed so far is then something like this. Scientists are grouped in universities, and identified in other research institutions, according to recognized scientific disciplines that have developed historically along with the university system. Disciplines can be analyzed into subdisciplines, and then into research groups. Research takes place primarily within these small groups, which are aware of other groups working on the same or similar problems. The direction and style of their research is determined at least partly by what these other competing groups are doing. Within such groups, and between a small number of them, full exchange of scientific information is possible because of a shared feeling for the nature of the data, the available techniques, and so forth. This is therefore the crucible within which scientific progress takes place.

Scientists may communicate with scientists outside their primary research group, and such communication may be quite fruitful and provocative in the sense that it may give a scientist a view of what is happening elsewhere and provoke ideas of potential value for his or her own research. This casual communication is widespread in science, and quite loose, as when colleagues simply chat about what is going on in their respective fields.[24] From time to time, a new idea or technique of considerable promise will be developed by a member of a research group, and this will begin to draw the attention of other scientists. Increased communication between the scientists attracted to this idea and either recruitment of other scientists or the weakening of ties to them will create a cluster of scientists operating with this new idea, or paradigm. These scientists, of course, may be at various institutions, but they can collaborate on research, and they may actively recruit students or younger members of the profession to pursue similar lines of research. The new work will appear in some journal and trigger, if it is successful, a flurry of articles in increasing numbers that pursue the same themes. In time, a successful cluster may form the basis for the creation of a recognized specialty within a discipline.

A process of the kind just described bears some relationships to Kuhn's account of science, and some differences. The new paradigm need not arise from a revolutionary crisis and a break with older science, rather it will typically appear as a locus of new interest within a discipline already sharing various values, and its potential interest will be clear to at least some discipline members. After initial successes, the cluster need not give way to a revolutionary break. Rather, as a specialty, it may have great longevity within a field in spite of (permanently) recognized anomalies and difficulties. The rate of contributed literature will settle down, and perhaps decrease, after initial successes, but the sequence of Kuhnian revolutions need not occur. Rather, what happens is that good scientists become interested in a new project and soon switch to that, leaving second-rate scientists or cautious personalities in command of the older ship. The background context of a scientific structure within which paradigms can develop, flourish, and turn into recognized specialties is missing in Kuhn's account, since he chooses to notice homogeneity only within the paradigm groupings.

The way in which research groups form around topics of interest, become solidified into permanent specialties, and shift membership is not structurally different from the way in which human groups of all kinds form to attack problems, and it is not unique to science. In many cases, people who form groups have such disparate backgrounds that the problem to which the group is directed and certain values concerning possible solutions are the only common denominator of group members. Early promise or early successes will hold the group together, but the group may disband or achieve some steady-state structure as the problem is solved or is seen to require constant attention. A group of ordinary citizens forming within a city or town to address some topic of mutual concern may show this dynamic. What is different about a scientific research group is not its internal history in this sense, but the scientific background against which it arises and against which it dissolves or turns into a permanent feature. Progress is not so much consensus, or the enforced unity provided by a paradigm, since many human groups are unified in terms of goals and strategies, but the fact that research groups recognize their position in the framework of a scientific discipline, with which they share fundamental assumptions.

Few human institutions have been large enough, sufficiently well organized and divided (conceptually), to offer this opportunity for research groups to intensify in order to work on a problem of generally recognized significance given the background institution. The theological organization of the medieval churches perhaps provided such an opportunity, but with a difference. Science as an institution is polycentric, and as we have seen, it does not interpret fixed texts so much as it constructs and interprets new scientific texts through interaction with nature. Polycentrism is an important aspect of scientific structure. There is no doubt that the polycentric competition of universities and research institutions in Germany, England, and the United States has contributed to healthy science, and there is some suspicion that the relative lack of polycentric structure has hurt French science. Science is, of course, at the same time elitist. A few scientists make a disproportionate contribution to the literature, and only a few scientists can gain wide reputations outside their own scientific specialty and disciplinary contacts. Polycentrism allows this chance for fame to appear in a variety of places, so that every scientist is potentially close to the opportunity for achieving a wide reputation, a fact of some importance, as we have seen, for morale. Polycentric structure thus quite naturally replaces the struggle of individuals for recognition, the process that preceded institutionalization. It is possible to imagine motivation sustained in a social structure in which the status of scientist is one of the few opportunities for superior financial and political status, so there is probably no essential connection between healthy science and polycentrism, but polycentrism is a stable and successful structure for handling novelty and status when the associated and traditional journal structure, refereeing system, and so on, are free from too much corruption.[25] Although polycentric structure allows a variety of locations for success, it is important to realize that this is coupled with the fact that no single location is the determinant of scientific success. The various locations make proposals, and the prestigious locations have the best opportunity for success in the resulting process of valuation, but disciplinewide recognition comes from convincing a large number or a majority in the discipline that some theory is best, or that some experimental result is right.

Polycentrism allows research groups to intensify and proceed, and then make their appeal to scientists of widely different backgrounds and convictions. Provided communication is reliable, and success is recognition by others in open competition, the polycentric structure of science is stable and progressive. Now it is possible to imagine that the reward system might not be polycentric, that is, dependent on recognition by one's peers. Financial and political rewards might be given to scientists by some bureaucracy, and this is compatible with high motivation and good science in the long run, provided that the relevant political authority has the astonishing ability to recognize good science and to bias the award system to reward only the best work. Given what we have said about scientific history, however, such ability seems to be nothing more than a bureaucrat's illusion. The nature of developing significance in science seems to require competition, and then reward, as the polycentric system of reward operates. An effort to take the waste out of the competition must require a conception of methodology that allows one to calculate the future significance of scientific work in the present. Because of this, the threat to polycentrism posed by the funding of big science has serious consequences for the dynamics of scientific progress.

To this point, we have looked at similarities between disciplines, subdisciplines, and research groups. We have found, tentatively, that all scientific disciplines can be analyzed into small research groups that have similar structural properties. When we look carefully at subdisciplines within particular disciplines, some interesting additional structure can emerge. Chemistry, for example, can be divided into eight subdisciplines, between which there are differences in consensus, instrumentation, competition, size of research groups, secretiveness, and so forth.[26] There is a tendency for subdisciplines emphasizing theoretical work to be represented by small research groups, and subdisciplines emphasizing applicability to be represented by larger research groups. This fact does not seem surprising, since the instrument of theory is still frequently the pen or pencil, but the instruments of applied science are likely to require special personnel, and the results likely to benefit in significance from repetition and from a variety of viewpoints. A subdiscipline like organic chemistry uses a variety of spectrometers, instruments that do not require special personnel. Spectrometers do not correlate to specific problems, since they can be used to characterize virtually all organic compounds. Organic chemistry, therefore, frequently exhibits small group or even individual research. In such an area, an individual scientist (with research assistants) may do a lot of experiments, publish many papers compared to a theorist, but produce work that is not widely cited directly by others, since it will be picked up into data compilations if it is of high quality. A theorist will publish fewer but more widely cited papers if he or she does good work, and a biochemist may work on more varied topics within larger research groups. This suffices to introduce the concept that research group size and style may depend on a wide variety of cognitive and experimental factors. We may expect to find correlates to these differences in chemistry in the subdisciplines of other scientific fields, so that while such research may tell us something about research within subdisciplines, it will not reveal much about differences between disciplines.

Having failed to find interesting differences between scientific disciplines in an analysis of the social structure of research groups or subdisciplines, we can return to intuitions about the differences in the disciplines as one moves from physics to sociology along the sequence of traditional disciplines. Many of them, of course, fail of demonstration. As we have noted, citation analysis found no differences in the temporal pattern of citations in physics and sociology although the intuition that there has been greater progress in physics seemed to suggest that physics citations would, on average, refer to more recent literature than sociology citations. As we move from physics to sociology, the number of normally recognized subdisciplines or research topics seems at first to increase (chemistry and biomedicine) and then to decrease again.[27] Even when the number of scientists in the disciplines is considered, this fact seems to say little about consensus or progress in these fields. There is a tendency for group work to become less frequent as one moves from physics to sociology, but the factor of instrumentation makes it clear that the existence of group research can be quite independent of considerations of consensus or progress. In other words, particle physics experimentation requires group effort to operate the relevant equipment, but sociologists can run surveys by themselves and analyze the results on a computer. Instrumentation may also explain why the cost of research per scientist declines steadily from physicists to sociologists. Computer use is now common to all fields, but beyond this, instruments differ. Physicists use enormously expensive instruments, and in many cases highly specialized instruments, while sociologists use survey techniques and statistical models that are common to most sociology subdisciplines and do not depend on specialized instruments. In publishing, there is a tendency for physicists to publish more letters and articles than books, and for sociologists to publish somewhat fewer articles and more books.[28] It could be thought that this indicates greater consensus in physics, since the article may presuppose more, but this is hardly an obvious conclusion. Articles are designed for other specialists in research groups oriented to similar topics. In physics (and mathematics) an attempt to go outside the research group runs into the problem that the special languages of research in physics are not immediately intelligible to other researchers. Articles are written because only other specialists are the intended audience, and they can understand the article format because of shared expectations. In sociology, there are no specialized languages of research, at least none that can't be explained relatively quickly in an ordinary language. There is thus the permanent temptation to argue the significance of specialized research results with a wider audience. The small amount of time that must be invested to do this may pay great dividends in disciplinewide recognition, while the time investment in physics would be substantially higher, partly because of other research that must be mastered to advance a large thesis, and this time investment would *a fortiori* tend to cause the author to fall behind that state of research in his or her home discipline.

The difference between the division of labor in physics and sociology, for example, seems intuitively real, and deserves further study. In physics, research projects oriented toward some goal can be broken up in many cases into smaller problems, with research physicists separating into groups attacking these problems. Specificity of problems means that communication with those working on other problems is made difficult by shared assumptions that may be difficult to articulate, but where the division can be justified it can be assumed that a good solution to a subproblem will contribute to a solution to the larger problem.[29] A trained physicist learns to convert his learned skills into instruments for attack on the specific problem that is the focus of his research group. By contrast, research problems in biomedicine or sociology are likely to be oriented toward projects that are not so divisible, and with respect to which a solution to a subproblem may cause great hazard for the possibility of solving another subproblem. Lowering the death rate in one area may raise it alarmingly somewhere else just through the technique that caused it to be lowered in a special case in the first place. Biomedicine and sociology seem to deal with systems that are not so easily broken into independent subsystems. Thus researchers may tend to bring existing expertise to bear on the problems involved, and may feel more of a necessity to communicate with others working on the same problem in order to avoid overlooking consequences of their own line of thinking. While these differences may tend to appear, and while typical physics research projects may be differentiated from typical sociology research projects using structural properties related to these differences, it seems unlikely that physics and sociology can differ uniformly in this way. Physicists can study systems, and sociologists have some divisible problems. Research strategies are therefore unlikely to be specific to disciplines, although there may be tendencies for disciplines to be associated with research projects utilizing recognized strategies.

Differences between physics and sociology can perhaps better be traced to differences in the way in which their problems arise and to differences in the way in which their theoretical vocabularies take on significance. For sociology, significance in theory is generally related to significance that has already been determined by everyday experience. Sociological theory, by comparison to physical theory, uses explanatory variables that have relevance to everyday life, for example, income, profession, age, sex, status, religion, race, residence (rural or urban), political attitude, and social class.[30] Social class is clearly the most theoretical term here, and this may be why the notion has been resisted by empirical sociologists. Some Marxists and various other social theorists have suggested that the failure of empirical sociology to develop theory like that of physics results from a concentration on the ideological veil of society as it is given in ordinary language, a concentration on concepts that deliberately mystify the actual structure of a society. This critique, while in many ways cogent, doesn't seem to go directly to an even deeper problem.

Physics deals with the constituents of very small systems up to larger molecules, and in its experimentation it is free to handle large aggregates of theoretically identical objects of scientific interest (for example, electrons) that can be prepared for experimentation and brought into a (theoretically) identical state. When human beings are objects of consideration, as in psychology and sociology, it is clear from theoretical biology that there is a problem.[31] Any pair of human beings will almost certainly differ genetically, and if not genetically, they will theoretically differ because of differing past experiences. Any two human beings thus have the theoretical potential to behave differently in any circumstances where their biochemical or cognitive capacities may make a difference. In physics, experiments can be replicated, bugs ironed out, and experimental results tied down definitively. Two physicists can know that their similar instruments should interact with interchangeable experimental populations in the same way. Experiments may be repeated on theoretically identical populations. A physicist may therefore freely project from one experimental population to all similar populations, and he or she may regard the experimental population as a fair sample from a much larger population that is well defined. This has considerable methodological impact in justifying various inductive policies and rational expectations. The experimental situation in physics underlies a well-defined division of labor. The theorist may know that the data of the experimentalist are exactly what the theorist would have obtained if the theorist had experimented in the same way. The theorist is therefore free to theorize without worrying about the legitimacy of the data used to constrain his or her theoretical development. In physics, so to speak, the experiments are the fixed points, and theorizing is the luxury permitted to physicists with respect to their data. A physicist obtaining data at odds with previous experiment, or at odds with legitimate theoretical extrapolation from previous experiment, will be cautious. Sources of error will be canvassed, and the experiment redone. The risk of publishing variant data is that someone else may do a better experiment, exposing the source of error and diminishing the reputation of the overly hasty scientist. In sociology, theories are treated as fixed points, at least in the sense that they can be understood without the relevant experimental data. A sociological survey of college freshmen in California produces a certain result. If another sociologist gets different results in a survey of sophomores in Illinois, the researcher is free to publish without the same worries. The second survey is just as good as the first and may be just as reliable, since there are many relevant differences that might explain why two populations would react differently. Further, the interaction of a sociologist with his or her subjects is subtle, and may influence the results, so in fact there can be only an initial presumption that the two experiments were the same. The sociological theorist is not therefore able to assume that personal experimentation would lead to the same experimental results reported by others in the literature, and for the reason noted, a sociological experiment can never be exactly duplicated, a source of dubiety for the theorist.[32] A sociologist with variant results may rush into publication without the same level of anxiety of a physicist. Sociology thus tends to produce a pattern in which the lack of control possibilities for determining repetition and differing theoretical expectations is associated with the accumulation of divergent and even contradictory results, which can be weighted quite differently by different investigators, while physics seems to orient toward eventual consensus and truth, because its partial texts are more easily woven into a consistent whole by repetition and repair.

If one wishes to retain the intuition that physics and sociology differ, it seems misguided to look for this difference in terms of the social structure of the disciplines or the research groups. The difference also seems to elude the invocation of paradigms, or feelings about the relative intelligence of the scientists in the two fields. Physics differs from sociology because of substantive differences in their subject matter and in the instrumentation that can consequently be used in experimentation, and because understanding enters the dialectic of theory and experimentation at a different point in the two sciences. Neither human free will nor human unpredictability is required to mark out this difference. The Appendix will take up this point in order to survey more systematically differences said to exist between the natural sciences and the human sciences, in spite of their sharing of the basic scientific dialectic of theory and fact. At this point, the relevant consequence of the preceding discussion is that there is no feature of the social structure of different scientific disciplines indicating that they produce substantively different kinds of knowledge, or that the knowledge produced by some scientific disciplines is more objective than that produced by others. All of this tends to obscure any suggestion that there is an unequivocal basis for making judgments about the relative rate of progress in different areas of science. Rates of progress must be traced to the availability of appropriate data domains.

Controversy and Progress in Science

If the long years of training of scientists resulted in the absorption of the norms of scientific practice so that a scientific habitude resulted, as Kuhn has suggested, there would be no cognitive reason for controversy in science. Similarly, if scientific knowledge could be grounded in observation, as in the empiricist and rationalist traditions, there would again be no cognitive reason for controversy. For any of these views, prior acceptance of authority would suggest that the counsel of reason would be to pursue cooperative further investigation. Because of the presuppositions of both sociological description and philosophical epistemology, controversy has always seemed to define some human limitation in the practice of science. Controversy between scientists is to be construed on these models as an eruption of human emotion over human reason, or as an eruption of a personal desire for scientific recognition based on some psychological imbalance. Because scientists are human, it might be argued, controversy must enter into scientific advance. One partial exception to this view is represented by Popper, who sees competition among ideas as analogous to biological adaptation, with the survival of the fittest ideas a desirable consequence of competition among them.[33] Since Popper makes a place for criticism in the competition of ideas, it may be desirable here to distinguish *controversy* from mutual criticism or even intense debate. It is genuine controversy, emotions included, that will be the focus of our immediate concern. Nearly every extant view considers controversy, over and above debate and criticism, a mar on the rationality of scientific behavior. And yet controversy persists in science.[34] If science is to be even partly differentiated from nonscience as a bearer of rationality and objectivity, it would seem interesting to discover that controversy is not a mere blot on scientific history.

Although there are controversies in science, it is also characteristic of at least some scientific controversies that they get settled, at least provisionally and for lengthy periods of time. The controversy between steady-state cosmologists and other cosmologists, quite heated for a length of time, resulted in the capitulation of the steady-state theorists. Adverse evidence took its expected toll, and it is easy to point to a variety of occasions in scientific history when controversy has been more or less ended by adverse evidence. By contrast, philosophical and religious controversies have stretched over centuries without attaining a clear terminus. It has already been indicated that controversy in science is not rational after sufficient text from reality has been produced so that only one plausible interpretation of it is known. Because text can be generated, the possibility of settling scientific controversy always exists, at least in principle, even if the techniques for producing crucial text are not always known in the early stages of controversy. But this returns our thought to its previous resting point. If controversy can be resolved by the production of text, why should more than cooperative criticism and mutual investigation exist in science if it were not for irrational factors?

One way to find a common thread in the irrationality of controversy—a way to make it rational, so to speak—is to shift consideration of scientific controversy from a cognitive plane to one of social interests. Bourdieu has provided a brilliant sketch of how this may be done if we regard scientists, not as pursuing truth, but as pursuing scientific authority, the scientist's socially recognized capacity to speak and act legitimately in scientific matters.[35] On this model, a certain amount of scientific capital is to be shared by scientists, who are then engaged in a partly political power struggle to maximize their share. In this struggle, recognition must be won from one's peers, but such recognition is never given for merely cognitive reasons. Successful maneuvering in this struggle is a function of origins, schooling, choice of field, strategy of attack on problems in the field, and so on, and not simply a function of intelligence and the quality of one's work. Indeed the quality of work is not a fixed datum; rather, it is something that one must struggle to achieve by establishing that it is more worthy of recognition than the work offered to the community of specialists by one's rivals. As in all political struggles, there are relatively risk-free rewards to be obtained by good party members, and there are risky but great rewards potentially available to those who subvert the present party and replace it with a new alliance. Clearly scientists choose careers and strategies within careers, depending on their assessment of their accrued scientific authority and on an estimate of their best means of investing that authority in order to maintain it or even to augment it. That science is not a purely disinterested quest for truth is shown by the anxiety surrounding anticipation of one's work by others, some of the controversy in science must be the result of the attempt by scientists to amass scientific capital, but only a small step along this path of analysis seems to threaten any hope of locating a sufficient cognitive component to allow room for objectivity in science, or for any other than an instrumental rationality in scientific practice. It would once again be inconsistent with the variety of scientific practice to assume that all scientists have had the career ambitions of pirates.

An attempt to see both cognitive and social aspects in controversy can be traced to the undoubtedly correct assumption that each individual scientist will have his or her own cognitive system. Any new piece of information must consequently be fit into as many diverse cognitive structures as there are scientists who find it interesting, but the significance of this information will vary widely with the cognitive structures into which the information must be assimilated.[36] To consider a crude polarity, new experimental data may be assimilated unchanged into the cognitive systems of many experimentalists, but be assimilated merely as support or denial of more abstract information into the cognitive systems of theorists, who will accept that the data do or do not fit theoretical parameters, with tolerance being granted to experimental error. If scientists are highly intelligent, and if they are highly trained, it seems inevitable that they will possess highly individualized outlooks. A contrary assumption based on acculturation must considerably overestimate the effectiveness of teaching in the learning process. Textbooks written by different authors always present theory slightly differently, finding a slightly different center for basic understanding of the theory, and emphasizing slightly different consequences of the theory as its most revealing aspect. Bellone has argued that each theorist must translate data into theory using a private dictionary, and that each experimentalist must translate theory into experimental design using a private dictionary.[37] Diverse dictionaries will contain slightly divergent pieces of mathematical and scientific information, but these pieces of information will be weighed differently in the dictionaries according to philosophical presuppositions. Because the dictionaries contain philosophical weightings that may be influenced by, or expressive of, social, religious, aesthetic, or political opinions, Bellone argues that attempts to separate internal and external history must fail. This picture is surely correct, but it threatens once again to divide the history of science into individual biographies that are difficult to bring into a meaningful historical pattern. What must in the end unite different theorists into consideration of the same theory is an orientation toward a certain range of data gathered in certain ways, and a recognition that the data constrain scientific thought along certain lines. The individuality of scientists explains why theories cannot be easily falsified by counterexamples, eliminated by data, or completely devastated by revolution. While a majority may revise their dictionaries drastically in the face of contrary evidence or the piling up of anomalies, some will work at conservative revisions, thus maintaining a wide potential variety of ideas in the gene pool of scientific adaptation to data. This point will take on increasing significance below. If the individuality of scientists explains why scientists might engage in controversy, particularly in a setting where the accumulation of scientific authority is at stake, it does not yet provide reasons for considering controversy to be a sound cognitive norm, although it may be taken to reveal why controversy will exist in all disciplines, independently of sociological measures of relative theoretical consensus.

If methodology could actually play the role postulated for it by many philosophers, then a scientist following correct methodology could be assured that the resulting piece of scientific information would have to fit somewhere into scientific explanation. Scientists could be regarded as fitting together a giant jigsaw puzzle whose pieces were produced in accord with sound methodological directives. A pervasive theme of this discussion is that methodology in the relevant sense does not exist. When a new piece of scientific information is offered, it is not at first known whether it is a piece to be fit into the jigsaw puzzle or not, no matter by whom it is proposed or how it is obtained. In subsequent discussion, this view will be set into a conception of scientific history and a conception of scientific fact that seem required to make sense of extant scientific practice. For the moment, it will be assumed that it is possible that the significance of bits of scientific evidence is not necessarily given to inspection. Under these circumstances, why shouldn't the self-doubt of scientists be converted into an even belligerent defense of the significance of one's work? Such a psychological mechanism is readily imputed elsewhere, but it has been neglected in the study of science, perhaps because the imputation of rationality and objectivity to science makes one forget that the ultimate status of new information there may not be different from the status of new information elsewhere. The scientist wants recognition that his or her discovery belongs to scientific history; thus the tendency to defend one's own claim takes precedence over the tendency to attack another's claim as unscientific. Should attack prove successful, it would still remain open whether the defended piece of information was genuinely scientific. What is being suggested here is that the open significance of new information provides a cognitive role for controversy. Controversy is one aspect of the struggle to establish that some new piece of scientific information is significant, that is, that it deserves to play a role in the thinking of other scientists engaged with the same problems.

It would be wrong in terms of the norms of public debate to simply assert one's opinion that some rival piece of scientific information was not significant. One problem is that at any given time in scientific history, apparently contradictory pieces of information may all be judged significant, the hope being that a novel way of fitting these pieces together can be found, or that they can be reinterpreted as noncontradictory in a new theoretical setting, as happened with the pieces of information put together into the quantum theory. The first cognitive priority is thus defense of one's own contributions. Other contributions can be attacked as self-destructing or careless. A contribution is self-destructing if it entails consequences at odds with its own assumptions and presuppositions. It is careless if it is not rigorous, either in logical development or in failing to note properties of experimental setups that are known to be relevant. Failure of an experiment to replicate shows carelessness of the latter sort may be involved. Where these strategies cannot be exploited, simply ignoring apparently rival claims may thus be an effective, energy-saving method of dealing with them that is consistent with the norms of public debate. This is a hostile indifference, and part of controversy. It is not unknown in science to have research areas fragmented into research groups that exercise only incestuous citation, simply failing to acknowledge rival work whose existence is all too well known.

If controversy plays a cognitive role in settling the significance of new work, what prevents controversy from destroying science, and from eating up the time that could be spent on other forms of productive practice? The contrary tendency is differentiation of interests and research goals. Differentiation can be used to avoid outright competition and its associated controversy. As scientific text is elaborated, it raises new problems, new questions, and new issues of interpretation. Rather than run the risk of losing in the competitive struggle to obtain some recognized desideratum, scientists may choose to engage a fresh topic where the chances of anticipation are minimized. Competition can be intense where there is consensus on important problems and there are widely accepted skills and instruments for working on the problem.[38] Rather than be crushed in such a race, some may prefer to choose a somewhat more private goal. Gilbert, investigating radar meteor research in England from 1945 to 1960, discovered more topics for investigation of researchers, and he also noted that the distribution of researchers across topics implied that researchers were not afraid of being anticipated in their current research.[39] At the same time, none of these lines of research was likely to cause a fundamental change in theory, or to promote wide scientific recognition for its successful completion. If scientists seeking a large increase in scientific authority may collide in the rush to establish priority in the solution of what are recognized to be important questions, in other areas of advance matters are more peaceful. Scientists can divide work amicably enough when more modest shares of capital are involved.

Controversy and differentiation can play cognitive roles in establishing the significance of new information and in finding new topics for research in the setting of an institutionalized science in which routes of scientific communication are well established. Because of this, what has been said here about controversy and differentiation is true of established nineteenth- and twentieth-century sciences, but requires modification for nascent science as well as science in the seventeenth and eighteenth centuries. There is no doubt that modern science began with a revolution in the course of which nature (as opposed to man) was secularized and became dead to man, that is, became an object for dispassionate scientific study. At one point, the implications of science for a world view were discussed at all social and intellectual levels, in church sermons, and so forth.[40] Before long, physics and astronomy had become autonomous from surrounding society. Later, biology passed through such a period in connection with evolutionary theory before it became autonomous. The universe of science is a bundle of laws and connections there to be exploited by man. Statements about quasars, plasma, and superconductivity do not excite public debate. Even nuclear technology and recombinant DNA research, although issues of public concern, do not seem to threaten a change in world view so much as termination of the world, period. The technical issues require the injection of expert testimony, contradictory as it is. Revolutions in thought seem possible when new scientific fields are carved out, and they have occurred in the past as the process of establishing scientific disciplines has taken place. But once a field is carved out, it becomes autonomous or nearly so, and discussion of the sort that can influence expert opinion in the field becomes a matter of expertise. Ties to the general intellectual culture of surrounding society are lost.

We will consider, for our purposes, only the autonomous research groups of contemporary science. Given what has been said about controversy and differentiation within such groups, it is clear that the chance of communicating with others and winning recognition is dependent on advancing information that is easily intelligible and readily assimilated. This fact encourages small steps, the sort that fit into an existing pattern of thought, and hence can be scrutinized from the many perspectives of other researchers. Further, the high cost of entry into modern research in terms of schooling means that while individualized outlook will be the rule, there will be a background of acquired scientific plausibility whose transgression means that work will be ignored or simply labeled as nonscience. The sum and substance of this observation is that revolutions in the intuitive sense of the word are probably nearly impossible in established sciences. Small steps and differentiation of existing specialty will be the rule. For this process, mathematics and formalization are helpful, since they allow comparison of the differences of theoretical frameworks similar enough to be brought into mathematical or logical comparison. Revolutionary activity, or genuine attempts to subvert the existing outlook, would have to depend on the expenditure of a good deal of amassed scientific authority if they are to have a chance. Bohm's attempt to challenge quantum orthodoxy is interesting in this connection. Without his prior reputation, his attempt probably wouldn't have been discussed by other scientists, and it is interesting how a presumed proof by von Neumann that hidden variable interpretations were impossible was used by many physicists to lay Bohm's efforts to one side without discussion.[41] Genuine revolutions must be rare in the research groups of modern science.

In spite of the abstract arguments against the occurrence of revolution in mature science, Kuhn's influence seems to have led to the search for revolutions in science. The theory of plate tectonics can be described as a revolution in geology, or one might speak of the Keynesian revolution in economics. There are periods in which the sciences—indeed all intellectual endeavors—undergo rapid changes in outlook. At any given time, some disciplines will seem to undergo more rapid changes than others. Physics in the twentieth century, for example, seems to have undergone some rapid changes early on, and then economics, and then later biology and linguistics underwent rapid changes while physics seemed rather to be mapping out consequences of its earlier upheaval. Relative growth may seem like revolution, and if history is written in the right time scale, a rapid series of small steps from one point to another may be viewed retrospectively as a sudden jump from one point to another. In order to bring talk of revolutions into consonance with the abstract arguments that seem to favor small-scale changes, it is possible to argue that what are considered revolutions in established sciences are typically historical artifacts. An established body of techniques, instruments, theories, and data suddenly takes on a new significance and permits new linkages because of a seemingly modest new piece of information or a new theoretical outlook. This can be compared to the problem of speciation in Neo-Darwinian theory. If all viable current change consists of small mutations in existing matter that are modest enough to be compatible with interbreeding, one still has to deal with the existence of quite diverse forms of life (on the assumption of development from a small amount of similar original material). The solution is to argue that environmental channels can steer lucky mutations into new niches where the colonizing forms can gradually establish permanence in isolation from the point of origin. When great men of science are studied in detail, it frequently turns out that teachers and acquaintances who did not establish scientific authority for themselves provided a framework in which a small change led to an important difference. As history is generally written without this detail, it may seem that the great scientist accomplished much more than was actually possible.

The difference between history, in which revolutions occur, and research, in which they must be infrequent at best, can be seen in a microcosm in the research report. Since the research report is a cultural product designed to play a role in the struggle for authority and to advance the claim of cognitive significance for the author's work, it does not pretend to capture the historical sequence of events. Authors of research papers may simply announce for this purpose that they had such and such a bright idea, and then go on to details of experiment that corroborated that idea. It is not necessary for them to specify the concatenation of odd factors, such as the availability of certain equipment, a chance remark of a friend, and an error, all of which may have played a role in the origin of the idea; nor need authors even be aware of these factors. Nonetheless, a study of such instances has frequently revealed that the bold step or bright idea involved such a concatenation.[42] Wittgenstein prefaced his *Philosophical Investigations* with a remark by Nestroy that progress always seems greater than it is.[43] No doubt scientists who have grappled for a long period of time with some problem will see a sudden advance as more revolutionary than it will be seen by historians who wish to fill in the relevant context. Nothing is here intended to impugn the intelligence that provides steps forward. What it may be important to establish for understanding science is that all steps must be small enough to involve backbreeding into accepted ideas, so that no truly revolutionary steps can occur, just a series of quick, small changes. What we have is an evolutionary development in which some steps seem more productive than others, and may be given the title *revolutionary* to distinguish this fact.

The history of science can be viewed as a search problem. Scientists look for observational text and theoretical text that will give mutual significance. Very simple observations about search may illuminate the scientific model. Let us consider a baby lost in some woods by careless parents. A search for the baby will proceed rationally if the area to be searched is divided up among the searchers. If there is varied terrain, it might seem rational to have swimmers search along river banks, and to have climbers comb any existent cliffs. As the search team increases in number, the areas to be searched can be subdivided. Although no mathematical theorems are relevant, it would clearly not be optimal for everyone to look at the same location. Suppose the baby is found. The reward, if any, goes only to the successful searcher, but the successful searcher could perhaps not have found the baby except for the division of search and the activity of the other searchers. Scientific research, though not coordinated from above as the baby search may be, also might not be successful except for the activity of all the scientists involved. The successful scientist searches an area not covered by the others, and his or her search of this area might not have occurred save for knowledge of what others were doing. Research has to be seen as a cooperative activity in which not all acquire fame, but in which the activities of all researchers may play a valuable role. Just as the difference between the successful searcher and the unsuccessful searcher may not be anything intrinsic to their search methods, but a matter of looking in the right (or lucky) place, so the difference between the successful and unsuccessful scientist may lie, not in the embodiment of different methodologies, but in the matter of looking in the right place or having tried the right combination. This is the social dimension of scientific knowledge that epistemology cannot capture without an associated social theory. Epistemology leaves the dynamics of progress untouched. We will develop a social theory of science further in the next chapter.

We can vary the search metaphor to yield an additional insight of some importance. Suppose an unknown territory is to be searched and mapped for annexation by a government. A preliminary search might best be undertaken by scouts, rather bold and rugged individuals who can rapidly canvass the territory, look for prominent features, and bring back information of considerable helpfulness to the more reserved map makers who must painstakingly survey the territory and bring it within the confines of cartographic representation. Perhaps it is not surprising that we find different kinds of scientists playing a role in scientific history. Their joint efforts may be required for success, and yet they may have difficulty in understanding one another. At one time, improving a success rate from 1 percent to 3 percent in some experimental setup may be the important way to advance understanding, and this may require a scientist whose major skills are mechanical and manipulative. At another time, some bold extrapolation to possible new data by theoretical conjecture may be required. It is generally recognized that physics contains experimentalists and theorists, two quite diverse roles, and analogues can be found in other scientific fields. If it be conceded that diversity of scientific personality and style can be a major contributor to the overall success of scientific research, then we have located another aspect of scientific social structure that eludes epistemology, and because the roles are not labeled by the participants in every case and may be switched during investigation by the same scientist, this aspect must also elude any merely descriptive sociology. A last word on controversy and differentiation may be that they encourage this important diversity under circumstances where otherwise the maintenance of a gene pool of alternative but valuable scientific ideas would be swamped by consensus.

An analogy between scientific progress and biological speciation has been introduced that will be of value in considering the vexed notion of scientific progress. In biological evolution, an unchanging environment is adapted to by species that may differentiate and adapt to various niches in order to fill out the available biological space with a stable configuration and distribution of forms. A changing environment will be met by genetic change in which various forms from the combinational possibilities will be tried for adaptive success in the new environment. This process will result in a new stabilization if the environment settles down once again. Is the history of biological species a history of progress, as opposed to mere change? Species disappear, new species appear, and species may fluctuate widely in their numbers over time. The environment may also cycle, seeming to repeat itself from time to time. It is the ability of species to produce different forms, to maintain genetic diversity, that allows them to change and adapt over time. Inflexible species will die out in cases of sufficient environmental change. Progress is an obscure notion here. What we have is a process that allows life to continue as the environment changes. An earlier form would not necessarily fit in the current environment, nor a current form fit in an earlier environment. Rather than progress, we have continuance. Progress in a philosophically rigid sense can only be defined as a continual closure over time toward some goal. When one has one thousand tiles to clean, one makes progress as one cleans tiles, assuming that the cleaned tiles do not become so dirty again during the process of cleaning that they must be redone. An early candidate for the goal of science was a complete and accurate description of the universe. When this goal has been set aside, as it has been here, progress toward such a goal no longer can function as the measure of scientific progress.

More recently, the progress of science has been taken to be identical with its ability to solve an increasing number of problems.[44] There are obvious problems with this proposal, even if it seems the only current possibility for avoiding relativism while acknowledging that no independent standard of ultimate truth is available for measuring epistemic progress. Even if science can solve more problems than ever before at each relevant point in time in its history, this need not be consonant with the intuitive notion of progress. Suppose, for example, that science throws up ten trivial problems each year for ten years, and also ten important problems, but solves only some of the trivial problems. One might not wish to say that science was progressing under these circumstances, especially if it had been solving important problems, since the existence of science, and civilization, might be progressively endangered by the inability of science to solve its new serious problems. Problem-solving accounts of progress must inevitably run into the problem that it is hard to individuate problems and assess their relative significance in a manner that permits problem-solving ability to measure progress. At the level of individual research programs, it is easily recognized that during any period of time, some of them will seem to be progressing toward their chosen short-term goals of research, others will seem stagnant, and others may seem to be retrogressing. Science as a whole, in all of its parts, is surely not continually progressing at all. A retrogressing or degenerating program may suddenly reverse its direction because of a new idea or the discovery of new goals, but it may also continue to degenerate. What we demonstrably have with modern science is simply continued life for three centuries, and the ability (so far) to adapt theoretically to experimental data. Many older scientific solutions to problems are no longer relevant, many simple theories have been replaced by complex theories over more recent data, and so forth. We cannot extrapolate from this past history of science to a successful future, project continued progress, whatever progress may mean.

If we view the paleontological record of a species, we may discover that it grew smaller in body size and then grew larger again, or that it lost and reacquired some measurable physical property. Viewed as a succession of forms, little can be said about why such changes occurred, only that they did occur. In order to understand and explain such changes, a knowledge of the environment is required. Perhaps the environment was cold, then hotter, then cold again, and a larger body size was more adaptive to the cold environment. Philosophers who have studied only the internal history of scientific theories are confronted with something analogous to the fossil record of forms without a corresponding record of the environment. In order to understand the succession of theories, one must take into consideration the data base to which these theories were attempting adaptation. New data, for example, could change the problems to be solved and set a theory back until a new form of the theory designed to adapt to the new data was available. Underlying the history of theory is the history of data text. What is cumulative in the history of science is the gradual refinement of scientific instruments once they are introduced until they produce data that seem to be robust in the face of further refinement. For some instruments, for example, the microscope and telescope, no theoretical limits to such refinement seem imminent. But certain objects studied by these instruments have remained in the data base since their discovery, and information about them has been gradually made more precise. A changing data environment for scientific theories is like a changing environment for biological species. Progress is not guaranteed, but theory contains adaptive measures that allow it flexibility in the face of such change. Of course these remarks freeze the dialectical interplay of theory and experiment by making it seem that data are fixed for theoretical adaptation, but it is important to see that concentration on internal theoretical change cannot by itself lead to any insight into progress, nor can concentration on the problems that such theories can solve. Rather than attempting to find a goal or property of science that assures progress, we will concentrate on the mechanisms for theoretical adjustment to data that have allowed scientists to adapt theoretically to the data environments that their instruments have located. While not ensuring progress, this perspective will break with the evolutionary analogy, for the sequence of scientific instruments will allow us to find a direction in scientific research that mediates the pessimism consequent to the sheer evolutionary analogy.

· 3 ·

SCIENCE AND NONSCIENCE

Science and Pseudoscience

In the last two chapters, we have considered some philosophical and sociological attempts to capture the essence of scientific practice. It has been argued that these attempts have failed, and this chapter will propose that there is good reason for this failure. The plausibility of such attempts seems to rest on abstraction from a selection of settled instances of good scientific practice. If one restricts oneself to some clearly described cases of sound science, say the important inferences of recognized scientific greats, then one may hope to find a common core of rationality in these instances of practice. What must be guarded against is the mere supposition that if this can be done, instances of pseudoscience or bad science will somehow differ from this characterization. When we look at the past, we can perhaps draw a line between good scientific practice and practice that failed. But this line is also equivalent to the line between that practice which has led to current practice and that practice which has not. The distinctions that we can locate in the past from the viewpoint of the present are likely merely to sort the past according to the present, but are perhaps not likely to be a representation of the past as it appeared to its participants. The supposition that we can understand what happened in the past on the basis of our current descriptions of the past must rest on a supposition that sound scientific practice doesn't change over time. If this latter supposition is wrong, and if the nature of science and scientific practice can change over time, then perhaps the past as interpreted in the present can't help us to understand the significance of the present, since it is compatible with this idea that the present is always, so to speak, partly opaque to itself. We can, for example, find distinctions in the fossil record between organisms that survived and those that did not. If we like, we can retrospectively designate those that survived as fitter than those that did not, and look for the common characteristics among the former. Such common characteristics cannot really explain surivival, however, unless they are fitted to a quite detailed evolutionary theory and a knowledge of relevant environments. But in spite of our knowledge of the past, the interactions between organisms at present and present environmental uncertainties prevent us from making confident predictions about the future of current organisms. Evaluating science is somewhat similar. When we look at the past, its evaluation is implicit in our vocabulary, and the record of success is relatively clear. When we look at the present, there is much in dispute. Data appear at the limits of instrumental reliability that are absolutely crucial, and current mathematical formulation may not seem appropriate to even accurately formulate the data that are being obtained. In these circumstances, confident predictions about the future of science transcend the capacity of philosophical rationality. This chapter will argue that the historical nature of science precludes the realization of anything like the logics of scientific research that have been proposed as a satisfactory characterization of sound scientific practice. The differentiating features of scientific epistemology lie partly in a social structure that is not captured in the logic of the practice of the individual scientist, and this social structure proves complex enough to require its own social theory. Normative philosophies of science based on consideration of the practice of individual scientists or of manifest features of their group practice cannot provide the general distinction between science and nonscience that they take to be a desideratum.

Let us begin with a consideration of parapsychology. Parapsychology is not regarded as a scientific discipline in many quarters, but how can it be ruled out as a scientific discipline? It is surely not the case that all orthodox scientists reject parapsychology, since many orthodox scientists have come to accept at least the possibility of paranormal phenomena, even though they may not themselves be engaged in research on the paranormal. Among those who are experimenting on such alleged paranormal phenomena as the psi phenomenon are many skeptics of parapsychology, who are experimenting primarily to refute the claims advanced by parapsychologists. These scientists may take themselves to be attempting to prove *scientifically* that the phenomena claimed by parapsychologists do not exist, but amount to experimental artifacts or are the consequence of outright fraud. Those who are convinced that there are valid paranormal phenomena include orthodox and unorthodox scientists on any intuitively reasonable assessment, and those who conduct experiments on alleged paranormal phenomena include those hoping for positive results and those hoping for negative results. Parapsychologists have attained university posts, and publish technical articles in their own journals and in some of the most reputable psychological journals.[1] In view of this fact, why should parapsychology be considered deviant by so many orthodox scientists? When we turn to methodological considerations, the puzzle does not diminish. The computers used in parapsychological research, including those used to provide random sequences of stimuli for experimental subjects, do not differ from ordinary scientific equipment. Experimental design in parapsychological research often is as sophisticated as the experimental designs found in orthodox psychology. There are bases for rejecting orthodox and unorthodox psychology as sciences, but there is hardly an obvious methodological criterion for drawing a distinction between them within general psychological investigation.

Criticisms of parapsychological research have not produced a scientifically respectable alternative theory to account for the experimental results. The critics may hint darkly of fraud, although fraud discovered in a legitimate field does not count against the legitimacy of the field, or they may suppose that there is always an explanation within the framework of orthodox science for nonfraudulent results. This latter claim is easily recognized as a mere expression of bias, or as a kind of metaphysical stance, and it is not a clearly articulate theory meeting any of the usual criteria for scientific acceptability. If the claims of parapsychology turned out to be true, then our present general psychological conception of human beings as complicated neurophysiological mechanisms might be wildly wrong. Because it conflicts with the current general psychological conception of human beings, parapsychology is bound to be threatening to those who have invested a great deal of intellectual capital in that conception. They are not likely to admit having been dead wrong at such a level unless confronted with overwhelming proof. A physicist not interested in psychology might be amused by this dispute, or open-minded about claims in other areas of science, but many psychologists will not be able to avoid commitment on this matter, and commitment is going to be related to the psychologist's feeling about the legitimacy of the current general psychological portrait of human beings. The rejection of parapsychology is thus related to its inconsistency with widely recognized current attitudes among psychologists about the validity of their general approach. Reactions like that against parapsychology tend to confirm the suspicion that common values rule over wider segments of the scientific community than the research groups. Parapsychology can't be ruled out in terms of methodological weaknesses, although orthodox scientists may attempt this to make it seem that their rejection of parapsychology is objective. The problem of parapsychology in seeking scientific acceptance is its clash with widespread opinions about the nature and limits of legitimate psychology, and not its current methodological status. Indeed, since discoveries occur in unexpected places and at unexpected times, who could prove that parapsychology will not one day become an orthodox science? What can be proved is that should that happen, vast revision in current thinking will be required—and one may therefore assign such a low subjective probability to the scientific future of parapsychology as to make it an unattractive area of research. Rationality cannot accomplish more.

The attempt to differentiate science quite sharply from nonscience is usually set against a dispute between science and something not considered to be science at all. Some might wish to rule parapsychology out completely as a possible science, and would welcome a rigid philosophical boundary around science. What is not so frequently considered is that if scientific practice inside major scientific disciplines breaks down into small research groups, some of these will strike off on the pursuit of schemes that will seem wild to their contemporaries in the same discipline. Some of these adventurous sallies will revolutionize science. A basic point is that many of the most revolutionary new theories introduced historically into any scientific disciplines were at first regarded as completely wrong by contemporaries. All of this suggests that we don't want a sharp boundary around what counts as science, or that such a boundary serves no really useful purpose. It is impossible to tell of fledgling disciplines whether they are or are not scientific, or whether they will or will not be absorbed into recognized disciplines, or even create new ones. Such judgments come later. Clearly, there is a difference between central scientific disciplines and nonscientific disciplines, and we want to be able to understand that difference even if we can't draw a clear borderline between them. This situation is like that of the difference between being alive and being dead. There are creatures that are clearly alive and creatures that are clearly dead, and the difference is of considerable interest to us in many cases. This difference is not obliterated by the fact that it is difficult to define *alive* and *dead* so as to provide a criterion for adjudicating all cases. The positivists wished to provide a normative paradigm of good scientific practice and to draw a clear boundary between science and nonscience. The latter was undoubtedly a mistake. A sound paradigm may measure the existence of science. Any discipline close enough to the paradigm, fitting it clearly enough, could then be regarded as a science. Some disciplines would have an ambiguous standard. This is what seems desirable and the only picture that is compatible with the opacity of the present, the idea that we can't adjudicate all of the relevant cases at a fixed point in time from a contemporary perspective.

What may at first seem surprising to those who are captured by the idea that a boundary between science and nonscience can be drawn in principle is that claims *within* orthodox scientific disciplines may be rejected by the majority within the discipline, and that this rejection may be defended by invoking dubious methodological criteria. The problems encountered concerning parapsychology and orthodox psychology may be played out entirely within an orthodox area of science. A scientist's work along orthodox lines may be rejected for reasons that seem merely to gloss rejection, and not to express legitimate causes for rejection. British physicist C. G. Barkla won a Nobel Prize in 1917 for his discovery of a set of X-ray emanations from atoms in the so-called K and L series.[2] Prior to his acceptance of the Nobel Prize, Barkla had also announced a J series of emanations, at first accepted by others on the strength of Barkla's reputation, but then regarded by most physicists as a derivative phenomenon that could be explained by the widely accepted Compton effect. Barkla's experiments, including those that had produced the K and L series, focused a powerful heterogeneous high-intensity beam of X-rays onto various atoms. The J series phenomena were a natural methodological extension of the earlier experiments at a higher beam intensity. Other physicists, including Compton, were attracted to a new instrument, the spectrometer, that produced X-ray beams that were of the same wave length, but much less intense. Barkla's and Compton's theories are not in direct opposition, since they were invoked to explain data deriving from different experimental arrangements. But in this dispute, Compton sided with orthodoxy in quantum theory (and helped to develop that orthodoxy), while Barkla opposed various features of the orthodox quantum theory. Barkla admitted to error on the original J series, but adopted a new view about related J phenomena, which were taken by him to be organic or emergent properties of heterogeneous radiation, properties that could not be discerned in the single wave-length spectrometer of the Compton experiments. Barkla's opposition to spectrometer experiments was methodologically subtle, centering around the idea that low-intensity experiments might fill in a certain amount of detail, but couldn't lead to fundamental new discoveries. Thus Barkla does not represent conservatism, even if this theorizing was in terms of an older language, but rather an attempt to go beyond what he saw as the instrumentalism of quantum theory to a more realistic and comprehensive physical theory. Nonetheless, by 1923, Barkla was literally alone as an exponent of the absorption experiments he had pioneered, and he is often pictured as an eccentric crank in his later years, even though he attained many honors and still published frequently. Papers were published allegedly refuting Barkla, or repeating his experiments with homogeneous rays and announcing the discovery that Barkla's data could not be obtained with such rays. The latter is hardly an objection, since Barkla regarded heterogeneous rather than homogeneous rays as a cause of the organic mode of data that he sought. Barkla placed sixteen advanced degree students for J phenomenon work between 1924 and 1945, even though most orthodox scientists thought the work to be nonsense.

What is interesting about this case is that it is a case where nonscience or pseudoscience as judged by peers develops within an orthodox scientific community and is accommodated within that community, even though it is recognized as deviant. Barkla's J phenomenon can't be ruled out as nonscience in terms of methodological criteria, since it was developed as a smooth methodological extension of Barkla's previous science. It just happened that most physicists took the alternative path of using the spectrometer. One can suspect that Barkla's organic theory violated a very general value of physics concerning the analytical approach, that is, the value of the correctness of analyzing something into its constituents, where possible, for the purposes of understanding. Thus most physicists probably regarded the homogeneous beams as a more basic and potentially useful exploratory tool than Barkla's heterogeneous ray. But this analytical preference can't be refuted by logical or even methodological objections alone. Therefore, our second case is like the first in suggesting a community of scientists who have internalized various values whose violation normally entails the sanction that work performed under the violation is regarded as nonscience, and may be largely ignored by other scientists.

An important aspect of the Barkla case is that one and the same scientist engaged in both orthodox and deviant science. It is hard to credit the possibility that he was schizophrenic, or that in some other way his mind was clear only when he was doing good science, but was confused when he was doing bad science. Such examples tend to show that good science can't be separated from bad science merely by the logic of situations. Paradigmatic values shared by some group seem to be involved in distinguishing good science from deviant science or even pseudoscience that is pursued according to sound methodological principles, that is, the same principles that seem to characterize good science. A second example may help to underscore the importance of this point. As is well known, Pasteur did early crystallographic work on tartaric acid, work designed to connect the crystal structure of these acids with their optical properties.[3] One acid that did not rotate polarized light, for example, he discovered to have symmetrical crystals, while another acid that did not rotate polarized light he discovered to have left- and right-handed forms, which when mixed in equal amounts in a solution seemed optically inert. These discoveries completed a research program that had been begun by the French scientist Biot some forty years before Pasteur's work. Pasteur's work depended on the microscope and the goniometer (used to observe crystalline forms), and was quite in line with the traditions that had studied, for example, the optical properties of quartz crystals. After Biot was convinced that Pasteur's work was correct, his support in the Academy of Sciences led to the rapid dissemination and acceptance of Pasteur's discoveries. But then Pasteur pushed on to greater generalization. He attempted to link crystalline form to properties of living things and even to the universe. He began to bathe plants in unnaturally polarized light, and used magnets to attempt crystalline distortion. Pasteur's new problems were deviant, and his methods unorthodox. But he was not mad. His fellow scientists, worried about the effects of such experimentation on his career, helped to move his interests in the direction of problems of fermentation, where he made discoveries of fundamental importance.

Clearly, although the existence of paradigmatic values is an important feature of scientific practice, and helps to explain the reaction of orthodox science to deviant science, the existence of such values alone will not differentiate good scientific work from deviant work or even sophisticated pseudoscientific efforts. The context of such values helps to explain why some science is regarded as deviant at a particular point in time, and to define orthodox practice. At the same time, it is possible to recognize scientific practice in the work of pioneer scientists who do not share the same paradigmatic values that are shared by other scientists in a recognized field of science. These scientists may be creating new fields of investigation, and their work need not be based on a perception of anomalies threatening some existing line of research. Galileo can furnish a sufficient example.[4] Groups sharing paradigmatic values are typical of modern science, as in Kuhn's conception, and perhaps the existence of such groups is required to deal with the distinction between orthodox and deviant science. At the same time, such groups do not need to exist for science to be practiced, for science was practiced early on in the absence of such groups and is practiced by many pioneer scientists as an individual activity. What the individual scientist within or outside such a group succeeds in doing is to narrow down complex and confusing data and observations into a manageable domain, often defined by the data that can be gathered by various instruments. Different individuals can then know that they are investigating the same data in the same manner, and scientists can recognize the relationship of new work to older work that had been done previously in the same domain. Science carves out, so to speak, something akin to immanent Platonic ideas or Aristotelian forms from the world and studies their instances. Physics and mathematics show this feature quite clearly. The same numbers and electrons have been available to all mathematicians and physicists for study. Recognized groups help to define common problems and techniques, but they are not necessary, even though they are pervasive in modern science and essential to the dynamics of its development. A pioneer scientist, once again, may propose important new scientific information without immediate group support. Because of this, mere perceived deviancy at some time, even deviancy perceived by some relevant group of orthodox scientists, cannot be taken as a criterion that the deviant scientific practice is not sound.

The move to science from nonscience or pseudoscience is the move to relatively simple questions put to nature through experiment that have widely recognizable answers. Scientific questions are designed to find out how nature works, what reality is like. And as Chapter 1 suggested, reality *is* like the picture that the sciences develop, although it is also like other pictures. The move to science may be perceived in the triumph of medicine over astrology.[5] Before 1700, it would have been difficult to defend the record of medicine over that of astrological practice, since medical treatments, for example, bleeding and purging, were frequently harmful. Astrological practice may actually have been more frequently beneficial because of its nonmanipulative therapeutic practice. Medicine, however, gradually organized and achieved fixed levels of practice, both of these achieved in England, for example, through the formation of the Royal College of Physicians and its sponsorship by the crown. Medicine gradually controlled its patients through its determination of which sorts of medical questions could be asked, and what sorts of answers were permissible. Medicine also developed instruments to assist in asking and answering questions about a patient's body independently of the patient's opinions. Astrology, on the other hand, was ad hoc and problem centered, responding to whatever questions were brought to the astrologers by those seeking them out. As the grandiose claim of astrology was that everything is determined by the stars, there was no reason for astrology to appeal to instrumental limitations in blocking lines of inquiry suggested by certain questions. And as the astrologers were an amorphous lot, consisting literally of all those calling themselves astrologers, fixed questions and agreed-upon answers to them were not developed by an astrological society, and patients could get quite contradictory information by consulting various astrologers utilizing divergent astrological systems. Medicine gradually came to dominate nature, seeking answers to aggressive medical questioning through experimentation, and attempting to control the course of nature through a manipulative therapeutic practice. The victory of medicine over astrology for the status of orthodoxy is symptomatic of the development of science in comparison to pseudoscience. Both science and pseudoscience may be theoretical, and both may experiment. Science, however, simplifies the nature and range of questions to be asked of nature in a direction that promises answers to the questions, answers that many may study through appropriate experimentation with orthodox instrumentation. Medicine can admit to being stymied by a wide range of apparent illnesses, but it has standard and recognizable diseases and standard treatments for them. Medicine, in other words, observes the course of a specific illness in a specific person and suggests a specific treatment. A course of study was developed that placed this practice within the orthodox educational system. Astrology seems to have been unable to make these adaptive moves. Its treatments and questions remained too general. Persons could be grouped together by the heavenly configurations at the moment of their birth and then treated alike, but the groups have seemed so heterogeneous that various astrological theories concerning such groups remain in a competition that cannot be settled by appeal to common paradigmatic values and an agreed-upon experimental and instrumental practice. Whatever scientific truths exist in astrological practice, if any, they remain obscured in a sea of conflicting seers' opinions.

It has been suggested here that orthodox science and deviant science cannot be differentiated by methodological criteria or by reference to a generalized group structure. What seems to be the case is that a scientific field is created when a limited set of questions about reality is formulated and can be put to nature through experimentation with the hopes of receiving definite and repeatable answers. Further, orthodox science develops over time in the direction of refinement of the questions and the experimentation for answering them, whereas deviant science seems over a period of time not to develop in this direction. The general problem for any absolute separation is that what is deviant at some time may develop later into orthodox science, particularly since nascent fields are so often perceived as deviant by existing orthodoxy. There is no general method for differentiating science and nonscience, or for branding deviant science as nonscience. We have to look for history to make these determinations, and that always both provisionally and from a point of view in historical time. Deviant science may grow into science, make discoveries that contribute to later science, or provoke the growth of genuine science. This fact cannot remain mute within any philosophy of science attempting to deal coherently with scientific progress.

Science and Society

We have seen that many philosophers of science have attempted to construct a sharp conceptual boundary between science and nonscience, where nonscience is interpreted as nonscientific intellectual activities. In this chapter we will consider some related boundaries postulated to exist between pure science and applied science and between science as an institution and other social institutions. Where nonscience was in question, the major point of a boundary seemed to be that only science could lead to knowledge, whereas nonscience gathered and organized mere prejudice or dealt with fictions. The effort to separate pure science from technology and societal influence is slightly different in tone. Here the effort seems to be that science gathers disinterested knowledge, or objective knowledge, whereas technology applies this pure knowledge to ends chosen by industry or by the interests of some segment of society. Whatever pure scientists discover need not be discounted for personal prejudice or for subservience to nonepistemological interests.

The philosophical attempt to distinguish pure and applied science is perhaps also related to a familiar philosophical predilection to favor pure thought, in other words, to a philosophical predilection to favor activities similar to philosophical activities. Philosophy in many of its historical forms has been dominated by a quest for certainty, a quest for permanent knowledge, and the philosophical picture of science tends to be of one piece with this attitude. According to some philosophers, pure scientists may be construed as calculating the values of various theoretical functions on data values obtained by careful experimentation. For these philosophers, the ideal scientist deals solely with facts and calculations, and cannot make any other than a computational error.[6] Good scientific practice then exists on a plane that is literally above criticism. Suppose, for example, that pure and applied science could not be neatly separated. Then pure scientists might be held socially responsible for harmful consequences of their scientific practice that might have been reasonably anticipated. Surely this would make scientists more cautious, and it could deflect the pure search for truth away from directions threatening lawsuits. In any event, argumentation would then transcend the realm of the pure investigation of fact, and begin to involve the notorious complexities and uncertainty of normative disputes. If the application of science helped to sustain a social system of disturbing inequalities and injustices, and even pure scientists played a role in the maintenance of this system, optimism concerning science would begin to run against some unpleasant realities. For philosophers, it may be tempting to argue that the unpleasant consequences of scientific knowledge are a measure of technological or political ignorance, and represent merely bad applications of a theoretically neutral pure knowledge. This has the consequence that philosophers can analyze logical structures within the domain of pure science without getting involved in controversial applications of science, and they can undertake this without grappling with the questions of social hegemony that inevitably play a major role in thinking about the value of societies in which big science currently plays a major role. Without retreating into an antiscience or antitechnology irrationalism, a cynic might observe that a resolute terminological decision to defend pure science against applied science can quiet a wavering philosophical conscience over time. In this discussion of science and technology, it will be suggested that the decision to put a wedge between science and technology, no matter what its other merits, has precluded an understanding of the dynamics of modern science, since modern scientific progress depends at least partly on the technology required to produce the scientific instruments capable of wresting data from reality, and is inconceivable apart from that technology.

At first glance, a fairly sharp distinction between science and technology can be made out. Before modern science, there were many technological discoveries that obviously could not have been dependent on prior scientific knowledge. Inventions like the spinning wheel, the loom, and even the steam engine did not depend on concepts derived from theoretical physics. Various problems and needs were solved, largely by visual comprehension and craft refinement, but also by a steady process of tinkering with slight improvements. The technological discovery is typically a physical object or thing, or the discovery of a process for producing such a thing, rights to the production and distribution of which can be owned by specific inventors through such devices as patent rights. By contrast, scientific achievements are not owned by persons, although persons get credit for making them; rather they become part of the intellectual resources of the entire scientific community through publication and other means of disseminating knowledge.[7] Typically, as we have seen, the scientist puts special questions to reality that are designed to permit clear, recognizable, and repeatable answers. The questions for technology seem to come from more mundane problems and needs. How can this be done more quickly, more effectively, or more cheaply? These are the questions for technology. Technological questions can usually be phrased in existing language, and they may be put to technologists in a modern society by employers of some kind who wish specific problems solved. It is clear that scientific knowledge will be used by contemporary technologists to solve such problems, so that modern technology does not develop independently of scientific knowledge, as technology did in earlier times. Nevertheless, it seems impossible to extend these initial observations into anything like a satisfactory general *distinction* between science and technology, because the two seem inevitably to overlap in the careers of various inventors. From an earlier era, Leonardo da Vinci provides an interesting example, but perhaps Thomas Edison will be a more accessible figure for making this point.

Edison's own self-image wavered considerably. By 1880, Edison thought of himself as a man of science whose achievements, such as the development of the first long-lasting electric light bulb, should be acknowledged on the standard scientific criterion of first publication.[8] In the same year, Edison began to publish a journal called *Science*, the predecessor of the familiar current scientific journal. Although Edison made some purely scientific discoveries and at one time thought that he had discovered a new force, which he called etheric force, his fame rests on various inventions, of which his lighting system is a well-known example. Even in Edison's time, it was not at all clear whether his proposed lighting system was science or nonscience (technology). Edison was invited to scientific meetings to read papers on the scientific aspects of his work, including the lighting system. At times, Edison preferred to keep various test results private, thus violating the norms of exchange of scientific knowledge and evincing the behavior of an inventor, rather than that of a pure scientist. Acceptance of Edison as a scientist enraged at least one prominent American physicist, Rowland, who clearly had Edison in mind in the following excerpt from an address to the 1883 meeting of the American Association for the Advancement of Science:

> The proper course of one in my position is to consider what must be done to create a science of physics in this country, rather than to call telegraphs, electric lights, and such conveniences, by the name of science. I do not wish to underrate the value of all these

> things; the progress of the world depends on them, and he is to be honored who cultivates them successfully. So also the cook who invents a new and palatable dish for the table benefits the world to a certain degree; yet we do not dignify him by the name of a chemist. And yet it is not an uncommon thing, especially in American newspapers, to have the applications of science confounded with pure science; and some obscure American who steals the ideas of some great mind of the past, and enriches himself by the application of the same to domestic uses, is often lauded above the great originator of the idea, who might have worked out hundreds of such applications, had his mind possessed the necessary element of vulgarity.[9]

The passage is sufficient to indicate that scientists, as well as philosophers, have felt compelled to defend pure science against vulgar and self-serving applications.

Whatever the value of the idea of pure science to the motivation of scientists who are engaged in fundamental research, it is still clear that the payoff from science for society comes through applications that help to master nature, and the payoff from science for those who employ scientists is this mastery, plus, in many cases, related methods of social control. It is clearly utopian to imagine that at this point a diminution in technological achievement might make our lives better. Technological inventiveness might be applied to problems of wider social significance, and the benefits of technological achievement might be more widely shared, but human life would undergo drastic changes for the worse if technological levels of achievement were to lessen suddenly. This can be seen clearly in the case of food production. Modern farming machines, fertilizers, and new hybrid species have combined to provide the possibility of feeding large populations through the work of a few.[10] Land and labor productivity have soared during this period, and in the United States this achievement has been reached through highly successful agricultural research. This research effort is interesting in that the scientists involved have been located near the farmers whom they were to help, and this decentralization seems to have had positive results. The scientists involved were able to become familiar with the problems of local farmers, and local farmers found technical advice easily obtainable. In this program, scientific research was not separated from technical demands and was not pure, in the sense that much of it was related from its inception to potential farming problems. Productivity in the American food industry is quite clearly related to the highly successful USDA research programs, and would be inconceivable without these programs. Separation of research and related technology may often be desirable to free research from any control other than the attraction to lines of research promising epistemological advance, but it is clear that science must ultimately be coupled to technological applications if it is to be supported within a social structure. Just how this coupling can be best effected, and how research and development funds may best be spent, is a topic not within the major interests of this work, although it is widely pursued by economists under the rubric of science policy.

The picture of agricultural technology suggests that one difficulty in the problem of separating science and technology is that the problem is considered primarily in the context of a society in which the roles of pure scientist and applied scientist have reached vocational separation. Intellectual discoveries in the domain of pure science are attributed to a discoverer, or a group of discoverers, but the results become ideally accessible to everyone through publication in learned journals. Practical discoveries are protected by patent rights and a legal system protecting the economic advantages of these rights. It can't be concluded that the features of this system prove that a sharp distinction between pure and applied science exists in general. In the case of Edison the line is obviously blurred. Modern research in warfare leads to pure and applied discoveries, but the distinction is once again blurred because all of the information is treated as a military secret and the normal device of publishing or seeking a patent does not apply.

So far, attempts to consider the relationships of pure to applied science have suggested that no sharp line of demarcation is possible. For our purposes, however, whether science and technology can be made conceptually distinct is somewhat secondary to the fact that modern science contains a major technological component in its instrumentarium. Modern science is inconceivable without accurate pictorial representation and without the highly specialized scientific instruments that scientists use to investigate nature. This fact makes the question of the independence of science and technology moot, even though those concerned to defend pure science have frequently not noticed this important fact. It would be possible to write the history of science in terms of the instruments that have been available for scientific use. From this standpoint, it is not surprising that mathematics, astronomy, mechanics, and optics were among the scientific disciplines first to develop, since the data domains over which these disciplines theorize do not require complicated scientific instruments for their development. Of course, it is true that larger telescopes, and radio telescopes, for example, have extended astronomical knowledge, but the basic outlines of astronomy were set down with the aid of naked eye observation and the primitive light telescope.[11] Chemistry, requiring complicated and specialized instruments, developed later. It would be wrong to suppose that chemists might have discovered the structure of complicated molecules before the development of sophisticated equipment whose data could guide such discoveries.

In addition to the instrumentarium, philosophers have overlooked the importance of technological developments in the visual presentation of both instrumental design and factual material. Manufacture of scientific instruments would not be possible without exploded views of assembly, which manage to compress a great deal of information about their construction into an easily absorbed schematic form.[12] Visual thinking and visual metaphors have undoubtedly influenced scientific theorizing and even the notation of scientific fact, a point likely to be lost on philosophers who regard the products of science as a body of statements, even if the statements have been guided by an intuition into the nature of things.[13] Could modern scientific work be at its current peak of development without visual presentations and reproductions of photographs, X-rays, chromatographs, and so forth? To consider a more specific example or two, would astronomy be possible at modern levels without time-lapse photography, or physics without bubble-chamber photographs? The answer seems clearly in the negative. Even if the data gathered by many instruments could be expressed in statement form, it seems that in numerous cases a coherent grasp of data requires technological achievements that weld technology and science into an inseparable unity as a means of obtaining knowledge about reality.

How is scientific knowledge related to other kinds of knowledge recognized in modern societies possessing a scientific component? Many philosophers have been tempted to drive a wedge between science and society, or at least to emphasize the independence of scientific knowledge from other forms of human knowledge. As with the relationship of science to technology, their purpose in doing this is to avoid the recurrent problems of relativism and bias that seem to threaten if scientific knowledge is at all tainted by nonscientific knowledge and nonscientific interests. Science is seen by these philosophers as determining its own ends and pursuing them with its own logic. If this separation were possible, then an internal history of science would be completely satisfactory. An internal history of science seeks to explain the history of science solely in terms of goals set by scientists to solve purely scientific problems, and procedures (usually cognitive) determined by scientists to maximize the changes of attaining such goals. Methodologists may seek to write internal history by giving a version of methodology, and then showing that some important segment of scientific history can be explained solely according to this methodological conception except for lapses and errors in the judgment of individual scientists that might have been avoided if the methodology had been followed more rigorously. Actually, this has been done for segments of scientific history, particularly for the conceptions of methodology attributed to Popper, Kuhn, and Lakatos.[14] Curiously, the same segments of scientific history can be made to fit any of these methodological schemes, showing that the slack between normative methodology and the details of history can be adjusted to fit various prejudices.[15] An external history of science would argue that the course of scientific history can only be explained by referring to opinions, events, and needs, that are not purely scientific and are rarely mentioned in the archives of scientific research, and it would argue that such material must be used in explaining the course of scientific history.

At first glance, the argument between those who feel that internal history will suffice for explaining the course of scientific history and those who feel that it will not seems inconclusive. Scientists are human beings who do live in a society and have human problems, and it would be absurd to suppose that they could ignore their background completely when they were doing science. Only the fear that this must distort scientific investigation and adulterate it with subjective elements could motivate the desire to deny the legitimacy of the claims of external history. At the same time, it is surprising how far internal history can go in providing a convincing account of the details of scientific history. The only way to reconcile these two observations is to take the position that science is relatively autonomous from its social setting, that is, that scientific development requires less recourse to external historical factors than the development of other social institutions. Nonetheless, a caution based on the sweep of scientific history is in order. From the seventeenth to the nineteenth centuries, science was pursued primarily by individuals who were free to choose lines of research on the basis of personal preference, and whose research could be funded from personal income or by a patron. For this sweep of two hundred years, the internal history of scientific development is nearly the whole story. But in the last century, science has become much less independent of its surrounding society. Scientific research involves costs that in turn require the acceptance or coercion of large populations in sharing these costs. Individual scientists have given way to corporate or governmental financing. The relative autonomy of scientific research from the society in which it takes place is therefore perforce weakening, and this raises a variety of questions that cannot be easily circumvented.[16] Perhaps the most important of these questions is whether science as the successful pursuit of pure knowledge can maintain a high rate of progress if its autonomy is considerably diminished.

Now let us look a little more carefully at some of the relevant historical points. For an internal historian of science, the inescapable impact of surrounding nonscientific values and suppositions may only hinder the development of science. A scientist may be unable to think immediately of the solution to some scientific problem because background presuppositions prevent formulation of a correct scientific answer. An internal history of the logic of the development can be written in which this event seems to show that actual scientific history is only delayed by scientifically irrelevant factors. In this way, the actual lack of autonomy in the real history may give rise to preference for an idealized internal history along with hopes that explicit methodology can help idealized behavior to occur more frequently. A useful example to reflect on in this connection is Kepler's introduction of the elliptical paths of the planets around the sun in his picture of the solar system. Kepler began in 1600 with the usual view that the paths of the planets were circular, a view that had to be abandoned because of the accumulating data to the contrary. Kepler did not turn to the ellipse as the natural generalization of the circle, as we would. In standard analytic geometry, a circle is a special case of an ellipse, and this is perhaps why the ellipse seems the next most complex closed curve in space after the circle to many people. Kepler turned to the ovoid instead. The ovoid plays almost no role in modern mathematics, is difficult to handle mathematically, and has almost no instances in physics. But the ovoid does have one important property in common with the circle, a property it does not have in common with the ellipse, and that is that it has only one focus. In a theological setting, where people were trying to describe God's creation, this is important. According to the principle of sufficient reason, God always acts rationally, that is, acts on the basis of some persuasive line of argument. If the planetary orbits are circles or ovoids, the plan of creation is rational, since a distinguished place is obvious for the sun at the focus of the orbits. If the planetary orbits are ellipses, the sun seems to be placed arbitrarily at one of the foci.

At a later time, when the connection between the inverse square law of gravitation and the elliptical orbits could be made mathematically clear, the background influence of theology could be squarely confronted and laid to one side. When Kepler was working, the impact of theological background information made the move to the ovoid seem necessary, and alternatives could not be easily formulated and evaluated. Kepler came to the ellipse as a mathematical aid in calculating ovoid orbits, and it was some time before he realized that the data actually fit the elliptical orbit better than the ovoid that he was trying to locate. Hanson cites this as an example of a necessary gestalt organization in scientific thinking that distinguishes great scientists from mediocre scientists.[17] Perhaps this is better taken as an example of how background knowledge can coerce even pure theoretical speculation in science.

At this point the methodologist is ready to argue that logic might have solved this quite nicely. Had Kepler formulated the logical alternatives, he might quickly have seen that the hypothesis of elliptical orbits was superior to that of ovoid orbits. The trouble with logic is that there are too many alternatives in logic, and that even the formulation of logical alternatives is dependent on the vocabulary available to the working scientist or logician at a specific time.[18] Scientists, as we have seen, rely on an intuition about what is relevant. In doing so, they will make mistakes, but they will keep alternatives to a reasonable and discussable few. If science is subject to background constraints, how are these constraints to be transcended? The answer seems to be that they cannot be transcended by mere logic, since logic will frequently produce an unmanageable catchall hypothesis as one alternative in order to provide a full set of alternative hypotheses, and logical intuition will usually not suggest the interesting alternative within this catchall hypothesis that is needed for scientific progress. Put another way, logic can enable us to formulate alternatives that have already occurred to us, but it doesn't produce new hypotheses.

We have already observed that pure science, perhaps because of a desire to differentiate itself from applied science, has frequently argued that its pursuits are disinterested and motivated only by pure epistemological considerations. Scientists themselves have attacked one another with the charge that research may be motivated in some cases by social interests, a charge that compels an attempt at rebuttal, given the social norms of scientific controversy. In the recent IQ debates, those supporting genetically based IQ differences between various groups of human beings have been challenged with the claim that their science is not pure, but is motivated out of racial prejudice and is designed to implement biased social programs.[19] Even though purely theoretical rebuttal of the methodology of much of this work seems sufficient to indicate that the inferences it makes from data are mistaken, the theoretical rebuttals have often been accompanied by nasty ad hominem remarks about the purpose of the research. Pure scientists like to consider their research independent of social motivation or social demands, lest their research slide into applied or technological categories. If research is applied or technological, it seems to take sides and to look toward a predetermined end. It is, in short, no longer free to pursue the pure logic of inquiry. But there is the following problem confronting any able scientist, namely, that a vast welter of possible lines of research stretches out from the present point of accomplishment. Which of these lines should be chosen to guide immediate research? A scientist may choose to work on a line of research that he or she feels may lead to personal success and acknowledgment of that success by other scientists, rather than on that line of research which seems to offer the greatest possible benefits for mankind. (It can be assumed that these lines will sometimes diverge.) This choice is personal and subjective, but it is not usually recognized as such, since it is concealed in the passive prose of scientific papers that cite a direction of research without indicating why quite different lines of research were not pursued. Again, a scientist might pursue a line of research out of some social commitment. The mere fact that a topic is chosen one way or another does not preclude that good science may result from its investigation. There is no necessity for a scientist who has chosen a goal to cheat or even lower his or her scientific standards, whether that goal be motivated partly from personal factors or from some social commitment.

An example that may serve to illustrate this process without involving contemporary controversy is the phrenology debate in Edinburgh between 1800 and 1830.[20] Phrenology stood opposed to the beliefs of the Scottish Enlightenment in proposing a nativist explanation for human behavior, and in seeking educational systems, penal systems, mental institutions, and even factories that might modify the innate dispositions that it sought to uncover in human beings. These dispositions were thought to be coded in specific areas of the brain. The phrenologists anticipated that the brain would be a complex organ whose specific parts served distinct mental functions. Abnormalities in the suborgans of the brain would lead to abnormalities of behavior. Neither the phrenologists nor the Scottish physicians and philosophers opposing them could directly correlate brain anatomy with skull shape or behavior, but the phrenologists supposed that the shape of the skull was caused by the underlying brain, and they attempted to correlate behavior and skull shape to establish their convictions, although anatomical dissection of brains was also involved. Although phrenology is no longer regarded as a science, the conviction of the phrenologists that the brain was a combination of specific organs was much closer to the modern view than the contrary conviction of their opponents. A huge dispute arose around the frontal sinuses. These sinuses could cause nonparallelism between the skull shape and underlying organs. The phrenologists minimized any nonparallelism and their opponents maximized it. As usual in such disputes, the two sides actually saw and drew slightly different sinuses, but current anatomy is not incompatible with the phrenologist's drawings of the sinuses. Needless to say, the phrenologists were attacked by licensed physicians because of the contamination of their knowledge by social goals. The social goals of the phrenologists are undeniable, but so is the fact that their anatomical work was important for the advance of medicine. An interesting difference between the phrenologists and the physicians is that the phrenologists were outsiders, in many cases foreign to Edinburgh or not part of the scientific and intellectual elite of Edinburgh. The social interests of the scientific elite are also evident in their research, but not quite so vulnerable to attack because defending their institutional privileges could be carried out without wildly upsetting a social structure in which those privileges had their place. A great deal of science is done without overtly threatening surrounding society, and histories of science tend to rely on the history of orthodox science, ignoring the social commitment of orthodox and unorthodox alike. But when we look at the details, we find that much of current science comes from historical controversies in which both sides had some sound scientific intuitions, and in which both sides made contributions to later scientific understanding. Perhaps it is clear then that good science can grow out of a quite interesting choice of a line of research, and that the attempt to deny that these interested choices play a role in science is nothing but a mystification designed to make science mysterious and incomprehensible to nonscientists and to support the notion that pure scientists are quite different from their fellow citizens.

It has just been suggested that a scientist may choose a line of research for many reasons other than its sheer epistemological potential, and that this need not preclude good scientific work. On this model, however, the scientist is still conceived of as an individual making a personal choice of research topic. In the past, when science was pursued by individuals, science was pursued almost independently of surrounding social chaos or coercion, since individuals were still free to make their own decisions. The twentieth-century interpenetration of government, science, and technology threatens this picture, and may even threaten the existence of successful science. Internal histories must remain ignorant of this threat, since they dismiss this interpretation and its associated funding complexity from the field of investigation. Popper has been alert to the problem, but he has argued that science is the model of a free and open society, and that it can function properly only when it is allowed autonomy and feedom.[21] He sees science as a delicate cultivated flower of Western civilization, a flower that is likely to die if its conditions for existence are changed very much. The difficulty with Popper's view is that science does not seem as free and open as it is on his normative characterization, and it is quite clear that science has flourished in other than free and open societies in Popper's sense. Popper sees science as necessary to solve problems facing society, and is worried that it may not solve problems rapidly enough in a transmuted form. Ravetz has contributed a brilliant and incisive analysis of the state of contemporary science, arguing that the system of criticism between competing scientists that has maintained quality control over scientific work and supported the reward system so necessary for scientific motivation (in this respect his framework is similar to Popper's) is breaking down, and that science shows signs of degenerating in quality and becoming a social system in which the political trade-offs of other social systems are becoming depressingly operative.[22] By contrast to Popper and Ravetz, Kuhn's treatment of scientific history allows no place for such worries, and seems to assume quite blandly that a functioning scientific work force is a natural part of any modern society. It seems clear that we have reached a time when the interaction of science, government, big business, and technology precludes any simple notion that a purely internal history of science can explain the current direction of scientific research. The philosophy of science thus needs to broaden its scope in order to consider how a negotiated autonomy for science can best be brought about so as to avoid the obvious potential for corruption in government and business funding for scientific research.

Public funds can be spent in obvious ways on social programs that will clearly alleviate human misery. And, of course, it is possible to imagine fewer public funds being spent. When public funds are spent on scientific research, a huge gamble is taken, no matter how the funds are dispensed on the research. Money can be spent in an area where some scientific solution seems desirable—but the solution may not come. And money can be spent under conditions that mean scientists are free to pursue their own ends—nothing may be discovered, or something enormously valuable might be discovered. Science policy attempts to look for rational guidelines for spending such money, but the obvious risks make a rational policy difficult to imagine. The situation is not like betting in connection with uncertainty, where the range of alternative outcomes and their chances of occurring cannot be reasonably canvassed. Because of these difficulties, the two pure positions are either that government should support science as much as possible while maintaining no control over its direction (the classic stance of the defenders of autonomy), or that science should fund only those lines of scientific research that have reasonable expectations of public benefit. Neither of these positions is ultimately defensible. The position of autonomy isn't defensible because not all that research scientists would like to do can be funded by public money, and because research programs may conflict in their experimental desires, so that some principle of selection of programs for funding is requisite. The social good position isn't defensible because it does threaten the health of science, and because reasonable expectation is so hard to assess. Compromise positions also have difficulties, as can be seen in connection with the finalization debate in Germany.

The finalization debate in Germany has its origins in an apparently innocuous extension of Kuhn's theory about the history of science proposed by some members of the Starnberg Max Planck Institute.[23] According to the Starnberg group, Kuhn's model of history is essentially correct, and scientific history can be written internally through the phase of normal science governed by paradigms. But it is possible for a discipline to reach a postparadigmatic stage in which theories are no longer tested and extended so much as applied. At this point, the orientation of theory development is no longer controlled by an internal logic, and the application of theory can be fixed by external goals. This position concedes that research planning isn't possible when research is 'pure,' that is, devoted to novel topics and shifting in accord with new discoveries. Revolutionary epochs and early paradigmatic stages of scientific disciplines thus elude rational control. But in the postparadigmatic stages of scientific disciplines, externally imposed goals can affect even the internal logic of scientific development. This position is distinct from the position of autonomy, because it denies that internal logic always determines the course of pure science, and is distinct from any form of Marxism, which argues that needs external to science always determine the course of scientific development. As a result, the Starnberg thesis has been attacked from left and right, each side primarily arguing that, on its terms, the Starnberg thesis is wrong. The real problem is that the Starnberg thesis is dangerously vague, since no definition or characterization of postparadigmatic science is offered, and this leads to the threat of political chaos in any attempt to implement the Starnberg thesis. Agricultural chemistry and chemical engineering can be given coherent external direction, but are these postparadigmatic sciences comparable to postparadigmatic physics or chemistry or biology? The obvious feeling that there is some difference between these groups of subjects in terms of cognitive goals highlights an example of the difficulties with the vagueness of the Starnberg notion of postparadigmatic science. Perhaps postparadigmatic sciences can be externally controlled toward socially chosen ends, but not all scientific disciplines at any given time are likely to be in this convenient stage. How are the non-postparadigmatic sciences to be funded when they require large public expenditures? We are back at square one, and the depressing suspicion intrudes that the important problems confronting science policy may have no rational solution.

As in the previous section, it has been argued here that there is no absolute wedge to be driven between science and technology, and none to be driven between science and society. Science and technology have become inextricably entangled, and scientists are human beings, which means that personal passion and bias will affect them as scientists and that pressure from surrounding society will inevitably influence their choice of relevant scientific work. The effort to drive these wedges is an effort to preserve the epistemological purity of genuine scientific practice. While all of this is true, the history of science can be written as an internal history far more successfully than the history of governments or religions. All that has been suggested so far is that the relative autonomy of science is not to be explained by the purity of motive of the individual scientist, or by alluding to a definition of pure science thought to be instanced in the best scientific practice. In order to advance beyond this point, some important aspects of scientific history must be considered.

Science and Common Sense

We have already suggested against claims for the autonomy of science that the relative autonomy that could once be associated with the individual researcher is no longer tenable as a picture of basic scientific research in the age of big science. Science and technology, science and background social factors, even science and pseudoscience, are impossible to separate analytically in conformity with what is known about scientific history. At the same time, it was suggested that science is probably the most autonomous social institution among all the nonautonomous social institutions, and the one social institution for which internal history can come close to an explanatory account. It has been argued that the scientist cannot reason informally, nor make naked eye observations, any better than many nonscientists. The general cultural outlook of the scientist will also have an impact on his or her choice of scientific research. None of these observations engages the fact that the scientist's scientific reasoning is aided or amplified by mathematical systems, that the scientist's scientific observations are aided by scientific instruments, and that the scientist's expression is constrained by a social system designed to criticize and discuss that expression as part of a competitive process with clear, even if somewhat informal, rules of struggle. These features mean that the scientist is pulled along paths of discovery that may well be repugnant to common sense, even his or her common sense, but forced nonetheless by the structure of scientific investigation. Although common sense itself will change over time, it is clear that at any given time, science and common sense can be in direct conflict. Moreover, this situation may be typical. Scientific knowledge seems frequently to involve a 'break' with common sense, and its apparent autonomy must be related to this important fact. Let us look at this more closely.

First, the objects of scientific investigation are rarely, if ever, *identical* with the objects of common-sense knowledge, even if the occurrence of the same names in scientific discourse and everyday language suggests some relation of reference. One may note that one-time staples of common-sense explanation, such as 'fire' or 'air,' often simply disappear in developed scientific discourse.[24] The facts about some particular fire or a particular body of air may be studied scientifically for some purpose, but 'air' or 'fire,' elements of an earlier cosmology, have no important place in contemporary science. Now let us consider common terms. The sun is an object of everyday comment and also of scientific investigation. It is clear that the properties of the sun as an object of informal observation and as a scientific object are quite distinct. Scientists consider the sun to be a point, or an enormous oblate spheroid, or whatever, depending on the scientific context in which it is being studied. Many philosophers have argued that the sun of everyday life is the same object as the sun of science. If the propositions about the sun in the two discourses have the same reference but the discourses are contradictory, then to restore consistency one discourse has to be translated into the other, or both discourses have to be translated into some neutral discourse. Using the ontology suggested in the first chapter, we find it more appropriate to say that the sun is a complex object that reveals a certain side to everyday discourse and quite a different side to scientific discourse, but that neither of these discourses is neutral or primary. Both are directed to different ends and have different purposes. Thus there is only one sun, but there can be no full description of it in any neutral discourse, so we can't say exactly and fully what one sun actually is. This does not preclude our being able to say fully and exactly what the sun is in some discourse for some particular purpose.

Historically, there was a time when scientists were studying the sun without thinking of it as distinct from the object of everyday discourse. Later, the discourses of science and everyday language are distinct, if not incompatible. Some philosophers have been tempted to suppose that the meaning of scientific discourse is to be understood by tracing its roots to the meaning of everyday discourse. This seems to reconstruct a historical process. Scientists began speaking everyday language and talked their way into a scientific manner of expression. In many cases, such as *sun* and *mass*, the same word occurs in both discourses. But unless one holds the extremely dubious position that all languages are intertranslatable, it is clear that we can learn other languages by starting in one language, discussing features of the other language such as the general sentence structure, the meaning of particular intertranslatable words, and wind up communicating in the new language, even though the two languages are not fully intertranslatable. This simply means that some growth or change occurred that transcends translation. Participation in a new form of life or a new culture suddenly causes a new manner of expression to snap into place. Kuhn has pointed out that graduate education plays this role. The fledgling scientist, by *doing* science, comes to participate in a scientific conceptual framework and an associated language. This is why science can't be exported through books, but needs to be exported through participation in scientific education, including laboratory practice.[25] Perhaps, then, entry to the meaning of a scientific discourse is to be understood in one way by tracing its development from everyday discourse, but this does not mean that the meaning of the scientific discourse is reducible to, or translatable into, everyday discourse.

The leap to a scientific notion has been brilliantly discussed by Bachelard.[26] One example he discusses extensively is the concept of mass, which on his analysis appears as five different concepts in its history. In its earliest, naive realist, form, mass is equivalent to sheer bulk. We desire the largest fruit, let us say, for our dessert. But the biggest of a kind is not necessarily the best. There may be hidden defects. From this observation of a possible delusion, a concept of mass as density was noted, and the scale came to operationalize this concept. Not the largest head of lettuce, but that weighing the most is what I should like for a fixed price. This concept of mass is positivist and operational. Both of these first two notions, perhaps separately learned in the history of many individuals, are prescientific. Newton represents the break to science when he sets force equal to mass times acceleration. Now mass is no longer a direct observable, but plays a dynamic role in a system of theoretical relationships. Special machines to measure forces and accelerations are developed by physicists, and mass is embodied in a special mathematical system that is separate from ordinary discourse, involving a special notation that one must learn to operate with. Mass is now relative to other features of an object, and becomes a quite complicated notion. Relativity again complicated the notion with the distinction between rest mass and mass in motion. In the mechanics of Dirac, a single object may have two masses, one of them negative. This total movement is marked most decisively by the break from mass measured by scales to the theoretical mass of Newton. At this point, there is a break that cannot be translated back into earlier discourse. That is why a scientist can learn classical mechanics, not by reconstructing history, but by learning the language of classical mechanics directly and then doing things with it, including the making of experimental checks on theoretical reasoning. According to Bachelard, the creative dreaming of the scientist occurs because of the layered concepts, like that of mass, which are the instruments of speculation rather than the instruments of experimental check. The internal structure of scientific concepts, quite contrary to the desired conceptual simplicity of much philosophizing about science, is a valuable component in fueling scientific progress. What is common to various scientists is the existence of the various subconcepts of a concept like mass, although the emphasis or profile of importance of these subconcepts will vary from one scientist to another. A particularly valuable exposition or discovery by a scientist will appear in a discourse flattened out to the consistent expression of only one concept, but the significance of the work will depend on the linkage and resulting suggestiveness of the subconcepts. This Bachelardian view helps to capture the complexity of scientific autonomy. Discourse can be autonomous, but it tends to be influenced where it is speculative by the archaeological structure of concepts.

There is usually no question that physics is a paradigm natural science. The development of a concept like that of mass consists in gradually 'bracketing' away layers of presupposition that are involved in everyday discourse. The more recent and most advanced physics, conceded everywhere to offer the deepest insights into reality, is the most self-contained and least dependent on everyday notions. Physics has evolved steadily away from everyday experience and everyday discourse. This is not surprising in the least. Everyday discourse is self-satisfied. At the everyday level, we can explain everything, perhaps by recourse to divinity or to classification of something as bizarre, but nonetheless we are pretty self-reliant in our mother tongue when confronted with the problems of daily life. Dissatisfaction with the obvious and suspicion of the easy answer underlie the drive of scientific thought. This is shown in a most engaging way by Ronchi in his history of optics.[27] At one time, areas of physics incorporating human perception were thought of as legitimate branches of science, for example, acoustics and optics. They remain, but their luster is considerably diminished. As Ronchi shows in connection with the theory of optics based on mathematical geometry, the human observer looking at images formed by mirrors with various kinds of surfaces will frequently see images other than where the theory predicts. For example, a human will frequently see an image *on* a mirror surface when theory places the image in front of or behind it. Human perception is constrained in various ways not anticipated by the theory because of knowledge of the probable size of such objects as human beings and candles. Optical experiments therefore do not necessarily reveal reality as opposed to human perception of reality, and optics is not really pure physics. This insight is related to Bachelard's dictum that reality comes at the end of science, and not at the level of everyday discourse. We begin with our internalized cultural reactions to the world. After an epistemological break to science, we may attain a description of reality itself, at least of one stable and calculable aspect of reality.

To this point, a semantic shift in the reference of terms along with the introduction of new terms has been cited as involved in the epistemological break to science. It is also true that syntactic differences between scientific discourse and everyday discourse can be noted.[28] For example, the use of past and present tense may not mark a temporal distinction in the ordinary sense, but indicate whether or not apparatus being described is thought of as temporary or permanent. "A cyclotron consists of . . ." is different in this respect from "A glass tube was inserted into the apparatus at point A." The past tense is appropriate for describing apparatus set up only to function in a particular experiment. A second example is the peculiar use of *given* in scientific discourse. Consider this example: Figure 9.5 shows how the vapor pressure of a given substance changes with the temperature.

The use of *a* or *any* instead of *given* in this sentence changes its meaning. *Given* functions here to mark out a definite substance arbitrarily chosen from a range of implied alternatives, a task seemingly frequently required in technical and scientific writing.

Unfortunately, there seems to be nothing in the semantic and syntactic shifts required by scientific discourse to indicate what is distinctive about scientific epistemology. Almost any move to highly specialized interests will be accompanied by semantic and syntactic shifts, although specialized interests need hardly be scientific. Sports writing, movie reviews, astrology columns, horse breeder's manuals, cookbooks, and so on, may all be written in a distinctive style that involves reference to features of the world not coded into everday discourse. All of these areas may involve an epistemological break with common sense, and it may not be possible to translate all of the relevant assertions back into everyday discourse. Everyday discourse may not contain the resources to define concepts that are related to specialized and invariable experience, so that words for these concepts involving extended and discriminate perception are best specially coined for communication among experts who are familiar with their specialized problems. Once a few of these terms have been developed, the language of specialty then has a life of its own. Perhaps lengthy and circuitous paraphrase can suggest the meaning of terms in the language of specialty, but there is no substitute in their mastery for the appropriate experience.

Specialized languages for specialized purposes may be relatively autonomous from the languages from which they were derived. Perhaps English and German, or some other pair of everyday languages, are not intertranslatable. What this may mean is that the languages are associated with quite different forms of life or cultural settings, so that the typical speaker of one language may use a word to describe an experience in a kind of setting that is not at all common for speakers of the other language. A word may thus exist in the first language and have no real equivalent in the second language. Could a word be added to the second language? Yes, but the point of so doing, or the success in so doing, remains obscure if the relevant experience is missing. The languages as used in religion or law may not be translatable because the relevant experiences in the two cultures in these areas are diverse. But specialized interests mean a focus of concentration. Two horse breeders, boxers, coaches, cooks, or whatever from two cultures may discover that they can communicate fully adequately just in the area of their common interests. Scientific discourse in various languages has not encountered the sorts of difficulties in translation notorious in the translation of poetry just because of the specialized interests and concerns of the scientists. It is not that scientific discourse is primarily declarative and that assertions can be more readily translated than metaphors, as it is sometimes supposed. It is rather that science concentrates on specialized problems and utilizes experiences that are not culture specific in the usual sense. Science, however, is a secondary culture, like many other specialized interests, and it filters out unique cultural experience because of shared activities, and can hence be produced as a (relatively) autonomous subculture in a very similar form in a variety of primary cultural settings.

To this point we have been operating almost entirely with an epistemological and sociological set of tools. While repudiating traditional forms of epistemology for their failure to deal adequately with scientific knowledge, and repudiating the claims of sociological inquiry to locate the distinctive features of scientific knowledge, we have only been able to offer a vaguely characterized dialectical conception as an alternative to these traditions. That science is reached by an acculturation process into scientific culture involving an epistemological break with common sense seems to be true, but it once again fails to distinguish science from acculturation into other specialized interests. From this point on, we will set the critique of existing programs into the background, and turn to the development of a more positive characterization of our dialectical conception of scientific growth. The first element we require can be obtained from an examination of the concept of scientific history that underlies both the epistemological stance so common in philosophy and the sociological stance involved in statistical investigation.

Science and History

When we look back at the history of science from the viewpoint of the present, it is often possible to discern which were the important moments contributing to the history of science, and which were merely dead-end speculations. The former moments are simply those that have had consequences that led to the present. The latter have no important consequences in the present. Any supposition that there is a methodological characterization that can separate progressive from nonprogressive moments in scientific history ignores a basic fact about the temporal perspective from which we must view a historical moment. Viewed retrospectively, a moment has a significance dependent on what has happened between that moment and the present. A contemporaneous moment cannot yet have this significance. Unless the future significance of what occurs at some moment can be determined at that moment, it is quite possible that it is impossible to tell at a moment in time which are the important scientific achievements; this can be known only retrospectively. Thus our knowledge of scientific history may not help us to fully evaluate contemporary science or to project the future significance of contemporary science. In this section, the relationship between history of science and the philosophy of science will be explored. Philosophy of science has typically assumed that a rather naive narrative account of scientific history is not only possible but the only possibility. When a more subtle account of scientific history is introduced, it should become clear why the dialectical attitude is necessary to understanding science. It should also become clear why neither scientific fact nor scientific theory can be set down as having fixed significance. That scientific fact cannot be set down in this way is important in transcending empiricism, but its importance is not evident until a better historical conception is at hand. The dialectical development of theory and experiment is internally related to the difficulty of evaluation of the present moment in science, a difficulty that may be maximized in some sense in scientific progress.

The most subtle barrier to the conception of history to be argued for is perhaps the failure to realize that any narrative account of scientific history, undertaken as it must be from some point in time, will utilize the scientific vocabulary of that time in order to attempt description of the past. Unfortunately, neutral description of the past in terms of contemporary vocabulary is not possible. Current scientific vocabulary is that vocabulary which has turned out to have high utility for current scientific description. This vocabulary contains within it an evaluative component that reveals what is now considered to be good science. If the scientific past is described in present vocabulary, we can tell from the description what is good science, that is, what has led to the present. Any attempt to circumvent this is to reintroduce the confusions of the past into the account. To make it clear what is being argued here, two chemistry experiments from the history of chemistry will be considered. At first, these experiments will be described from the viewpoint of their own time. What was a brown stuff of indeterminate nature will not at first be given its modern name, but simply be regarded as brown stuff. This is how we shall try to stay within the correct time framework. Both of these experiments seem to fit any reasonable description of sound experimental technique when viewed from the perspective of their own time. When these experiments are redescribed from the standpoint of the present, it will become clear which of them was progressive and which was not. The modern description locates the source of error in the one experiment, since now that we know what a chemical element is and which substances are chemical elements, the one experiment has to be seen as attempting to prove an element to be a compound (in modern terms), and hence is hopelessly wrong from our point of view. This historical exercise may prove sufficient to point to the unnoticed innocence involved in any idea that we can provide a simple narrative description of the past.

The two experiments to be described are chemical experiments, and they were performed during a period in chemistry in which there was intense investigation by chemists of what substances were chemical elements and what substances were chemical compounds.[29] These experiments were part of a sequence of such experiments, and the two chemists who performed them were aware of one another's work, and were in effect criticizing one another by performing these experiments. One chemist sought to prove that one of the substances in his experiment, which we will call *X*, was a compound composed of chemical element *Y* and another substance, and the other chemist sought to prove the chemical opposite with his experiment, namely, that *Y* was a chemical compound composed of chemical element *X* and another substance. The two experiments are therefore chemically contradictory, and one of them must be wrong, given the structure of modern chemistry. If the experiments are described in modern vocabulary, the wrong experiment will be obvious, since we now know which are the chemical elements and which are the compounds. At the time when the experiments were performed, there was no uniform chemical vocabulary (the substances used had names like calx and minium), and the two chemists had to exchange materials in a sequence of experiments to make sure that the *X* and *Y* they were using were in fact the same substances, so that their experiments were in actual conflict.

Now let us describe the two experiments in fairly abstract terms. The first experiment is designed to show that *X* is a compound by making it from *Y* and *Z*, through the application of heat. The second experiment is designed to show that *Y* is a compound by making it from *X* and *U*, through the application of heat. At a very abstract level, the designs are nearly identical. In the first experiment, liquid *Y* is placed in a retort connected to a quantity of gas *Z* collected over water in an inverted glass container. When the retort is heated, a solid *X* appears on the surface of *Y* and a quantity of the gas *Z* over the water clearly disappears. This experiment seems to make *X* from

Y and *Z*, apparently establishing that *X* is a compound and *Y* is an element. In the second experiment, a quantity of *X* is placed on a wooden boat on the surface of water over which the gas *U* has been collected in an inverted glass container. An application of heat brought about by concentrating the sun's rays on *X* with a magnifying lens suddenly causes the water to rise in the inverted container, *X* and *U* to disappear, and a quantity of *Y* to appear on the boat now at the top of the inverted container. This experiment seems to make *Y* from *X* and *U*, apparently establishing the proposition that *Y* is a compound and *X* is an element, a proposition that cannot be chemically reconciled with that established by the first experiment. It might be noted that both chemists accepted (correctly) the idea that the glass, water, wood, and methods of heating were chemically irrelevant to the outcomes of the experiments.

Both of the experiments described seem methodologically sound. Indeed they are virtually similar in design, since heat is used to create a new substance from two existing substances. The difference between them lies entirely in the fact that an erroneous chemical conclusion was drawn from one experiment. When the experiments are redescribed by identifying *Y* as mercury, *X* as mercuric oxide, *Z* as oxygen, and *U* as hydrogen, the apparent conflict is easily resolved. Actually, the gas in the container in the first experiment was ordinary air, and the oxygen is removed from the air in the experiment, but this does not affect the chemistry. The first experiment makes mercuric oxide from mercury and oxygen, and the second experiment makes mercury from mercuric oxide and hydrogen, and now one sees immediately what went wrong in the second experiment. Mercuric oxide and hydrogen and heat make mercury *and water*, a fact that becomes obvious when balanced equations corresponding to the experiments are written out. Lavoisier performed the first experiment, and Priestley the second. At this point there is a methodological temptation to suppose that close observation might have detected the water produced in the second experiment, so that it could have been seen at the time to have been in error. In fact the possibility of water production in the second experiment was discussed by the two men, and the experiments were redone with the gases contained over mercury, rather than over water. Some water was noticed in the second experiment. Priestley attributed this to water impurity in the mercuric oxide or the hydrogen that he was using. This guess, while at least partly wrong, was perfectly reasonable and couldn't have been ruled out as wrong at that time. The manufacture of chemicals was not exact enough, nor were balances exact enough, to distinguish the error in the second experiment. In our time, the availability of better chemicals and better instruments, plus the conservation principles and the widely verified system of chemistry that has grown out of Lavoisier's work, enables one to show the error in the second experiment. At the time, the experiments were the crux of controversy that was quite explicit, but not resolvable in terms of the techniques available then. This is a good illustration of a point made earlier, namely, that science at any time is partly dependent on the available technology for producing experimental equipment. The overriding point is that our temporal perspectives give these two experiments a completely different comparative significance that would be lost in any simple methodological discussion from a contemporary perspective.

Priestley was a supporter of the phlogiston theory, which has been roundly criticized as methodologically faulty by many philosophers and historians of chemistry. There is a persistent tendency to argue that the phlogiston theory was logically inconsistent, entailing such notions as that of negative weight.[30] Consider once again the equation from the second experiment stating that mercury is produced from mercuric oxide and hydrogen (phlogiston), and try to view it from an eighteenth-century perspective. It was discovered that when a metal was burned, the resulting oxide was heavier than the metal that had been burned. This suggests that the hydrogen driven out in the burning has "negative weight," that is, less than no weight at all—thus causing an indirect suspicion that some factor is missing in the account of the second experiment. Phlogiston theorists took this factor to be water, and argued that burning could only take place where free water was available to replace the phlogiston in the reaction. Their observation was that there always was plenty of free water around in the crucial experiments. This is to recognize the importance of water in the reaction, but it places the water on the wrong side of the equation. Only a better analytical chemistry and later experiments could show this supposition to be wrong, but the pressure to make the supposition shows that the phlogiston theorists were as anxious as their opponents to avoid discovered logical inconsistency.

When we look at the history of science, it may seem that observation was precise and that the progressive can be sharply distinguished from nonprogressive. The example of the Lavoisier-Priestley confrontation, in combination with many other similar examples that could be produced, should be sufficient to indicate that this is frequently a result of relabeling the past with the labels of the present. Philosophers of science have tended to be too naive about history, supposing that the security of the present would suffice for neutral description of the past by an appropriate methodological shift.[31] A scientific discovery, like many historical events of importance, makes a number of later events possible, events that could not occur without it. When one of these possibilities is actualized, it gives the discovery a significance it could not have had before this actualization. There is a sense in which we can only understand a discovery *after* its potential consequences are realized. The process in science of attributing discoveries to individuals is the way in which this process is acknowledged.

What is being said here applies equally to the level of scientific fact and of scientific theory. The significance of both is determined typically over time after an initial proposal. Should science be considered a matter of fact and theory, it must also be considered a matter of significant fact and significant theory. A fact, to become incorporated, must be of the sort that gives some theory significance, and a theory, to become incorporated, must be of the sort that gives at least some facts significance. Theory unassimilated to fact and fact unassimilated to theory will eventually drop out of science after a trial period. At any given time, there will be assimilated fact and theory in science, but also newly proposed facts and theories. Their significance and their longevity will depend on the development of a tissue of connections to existing fact and theory as well as fact and theory yet to be proposed. Mindless accumulation of data is, in the long run, irrelevant. That is why totally accurate observation of some succession of events at some arbitrarily chosen point is unlikely to be of scientific value, and is also unlikely to remain part of the scientific corpus. It is unlikely to call for a theoretical explanation sure to be of value in predicting the course of events elsewhere. We can summarize this line of thought by saying that science is interested in significant facts and significant theories, that is, facts and theories of mutual relevance, and not in the accumulation of mere theory or mere fact.

That a fact requires a theory to be significant, and to be retained, is shown by our chemical example. The two experiments were designed partly to show opposite facts: one the fact that *X* is a compound, and the other that *X* is an element. When the experiments were first performed, there was *ipso facto* evidence for both of these contradictory assertions. Both experiments and their results were part of the factual basis of developing chemistry. But what the experiments showed wasn't settled until a chemical theory was developed that could explain the results of both (and many other experiments) satisfactorily, while a rival theory showing the experiments to have an opposite significance couldn't be developed over the same range of data. Then it became a fact that one experiment was taken to have a fixed significance, and the other experiment was reinterpreted. After this, *both* experiments dropped out as insignificant. They no longer had anything to say to chemists. Their significance was exhausted. What they had to say became part of the vocabulary of chemistry, and the focus of chemistry was on new topics.

Another example may not prove superfluous. A frequently cited example is Mendel's work on inheritance, published in 1866 but not really influential until 1900.[32] Was Mendel so far ahead of his time that his discovery was too revolutionary for his contemporaries to grasp? Or is it the case rather that Mendel's work was of no particular significance until Darwin's theory of evolution had run into difficulties? To begin with, it should be explicitly noted that Darwin's theory was not published (1868) until after Mendel had completed his work. Further, Darwin's theory, while explaining evolutionary change over time, did not contain a description of a satisfactory mechanism for explaining the rate of such change, or the fact that some characteristics of organism did not disappear, but might suddenly reemerge. Darwin's own hypothesis of pangenesis was inadequate to this task. Mendel's rediscoverers (De Vries, among others) were searching for a mechanism of inheritance that could support Darwin's theory and replace the hypothesis of pangenesis. What they found in Mendel was decisive evidence for discontinuous evolution, evidence leading eventually to the theory of the gene. Mendel's observation of ratios now took on a significance that Mendel himself seems not to have been aware of at the time of his experimentation, and could not have been aware of unless he had anticipated Darwin's theory and its difficulties. What we now encounter as Mendel's experiments, with their associated significance for evolutionary theory and the mechanism of inheritance, is a historical reconstruction to be dated at least thirty years after Mendel's own publication. From our current point of view, Mendel observed a fact, but the fact that Mendel discovered did not become significant until 1900. At that point, it became a significant fact because of its relationship to evolutionary theory, and it has since remained an essential element of the factual basis for biological theory.

Just as the significance of fact may be revealed by the appearance of theory, the significance of theory may be revealed by the appearance of fact. Non-Euclidean geometry and noncommutative algebra, as is well known, provided a way of interpreting facts in newly emergent areas of relativity theory and quantum theory. In the crudest cases, a sufficiently bad theory may be recognized as having no significance by sufficiently recognized fact, and a theory may be recognized to have greater significance than before when it permits the deduction of a fact that is then confirmed by experiment. Between these cases, the range of significance of theory may be widened or lessened by discovery of its ability to integrate apparently unrelated facts, or by the discovery of facts seemingly closely related to the facts that it can explain for which no existent alternative explanation seems available. This part of the dialectical interaction needs no particular historical examples, since the evaluation of prior theory by newly discovered fact has played an important role in empiricist philosophy of science. It is, of course, being argued here that neither theory nor fact has precedence, and that the significance of both is under constant scrutiny in any healthy science.

In the Appendix we will confront some problems associated with differences in the development of the natural sciences and the social sciences that will mean that the account given here is primarily descriptive of the natural sciences. For the natural sciences, the development of experimentation does not involve the difficulties confronting social experimentation. But in both cases, it will remain true that the past of a science exists in its present, in that what has proved most significant and valuable has been assimilated into the current language and instrumentation of science. The present, however, does not contain sufficient information for us to grasp fully its significance. This view seems true of human history in general and of scientific history in particular. Once it is granted, it is no longer possible to retain the idea that a fully comprehensive methodology for science is a real philosophical possibility. On this view, well-known examples from the history of science become other than indicators of stupidity and ignorance. As Holton has pointed out in a penetrating paragraph related to this point, neither the work of Carnot nor that of Planck could be understood or appreciated until some time later.[33] It is doubtful that even great innovators such as these could fully grasp the significance of their own work, and its delayed acceptance by others can hardly be set down to their unwillingness to adapt to new ideas. To be sure, at some points in time a set of perceived anomalies given prevailing theory may make a new theoretical position widely appreciated immediately, or the success of some scientist in locating a much sought-after fact may be widely appreciated by his fellow researchers, but for many innovators significance must await the construction of related facts and theories.

Empiricism and rationalism have often assumed that we can describe contemporary events completely. Cartesianism begins with the assumption that knowledge can be read from sufficiently clear and distinct ideas, and empiricism often begins with the assumption that knowledge can be read from sufficiently clear and distinct experience. We have already examined the difficulties with extending these positions into comprehensive philosophies of science. It should now be clear that a view that narrative history of science is possible, that we can tell what is happening in science by describing fully what is happening, is a deep and pernicious constraint on the philosophy of science, even where the philosophy of science is willing to concede the inadequacies of rationalism and empiricism. Once this is replaced by a subtler historical conception, the appeal of the dialectical approach offered here is transparent. Empiricism and rationalism have proceeded by assuming either fact or theory to be fixed. In this way, they have captured part of what happens in science, but they cannot ultimately justify the assumption that their basis is adequately self-explanatory for the construction of methodology.

It is indicative of the scope of the problem confronted in this section that in some sense a narrative view of history is shared by Kuhn with his opponents. In a recent review of a book by students of Lakatos, designed to indicate how Lakatos's philosophy of science can be supported by history, Kuhn argues that "fitting the data" is a rather obscure notion in history as opposed to science.[34] Thus a resolute historian with a philosophical ax to grind can very well force the facts to accommodate his preconceptions. Kuhn also (and quite rightly) attacks the notion of "actual history" used by Lakatos as a myth about the history of science. The pool of all data relevant to history cannot be identified or surveyed. On the whole, then, Kuhn expects to find little of historical value in the history of science when it is written from an explicit philosophical perspective. Kuhn is quite explicit in his insistence that history cannot be intelligently written as a narrative chronological sequence of events, and yet he also feels that theoretical preconceptions should not be allowed to influence the perception of historical fact. What is a given for normal science, that theory influences the perception of fact, is not to be carried over to Kuhn's work as a historian. In history, the facts are said rather to be capable of speaking for themselves, provided that preconceptions are set aside so that the facts can be heard.[35] At least for history, then, some traditional epistemological stance may turn out to be correct. Without endeavoring to describe how history can best be written, and without disagreeing with Kuhn's remarks about philosophical history, it is still possible to suggest that the facts in the history of science do not speak for themselves. Rather, they must speak in consort with other facts (theoretical or experimental) if they are to say anything, and they will change in significance and relationships over time, or be lost from the historical narrative and cease to influence current vocabulary and perception.

The view of scientific history we have developed now makes it clear why science cannot be sharply separated from nonscience in general. At any point in contemporary scientific history, there will exist clear items of science and research groups whose efforts clearly fall within science. Yet at the same time, there will be nascent lines of research whose evaluation remains for the future. While some areas can clearly be regarded as science, any effort to sharply separate science from nonscience requires that the significance of new work, its more permanent place in science, be assessed at the time that it is proposed. Any attempt to draw such a line by convention must inevitably lose the historical dimension of science, a dimension that is essential to its philosophical comprehension. The extent of new theory, its ability to accommodate data, will be determined over time, as well as the significance of data, their importance in adjudicating between theories. Data text in particular can be indefinitely expanded through interaction with reality. This feature of science is the major way in which it differs from nonscience, and it explains why the process of evaluation in science is never completed. The significance of data text becomes clearer as its context is increased through experiment. It may not last as significant in scientific practice, or its significance may be coded into scientific language or into scientific instruments. But a later text may suddenly shift the significance of an earlier one, and there can be no time limit on this process.

· 4 ·

SCIENTIFIC FACTS AND SCIENTIFIC THEORIES

The Social Construction of Scientific Fact

In earlier discussion, it has been observed that observation and thought at the growing edge of science are obscure and difficult to assess, quite in contrast to all forms of epistemological outlook, which rely on hard data, on the idea that nature's properties can conveniently be read in observation or experiment. By the time that experimental results are written up in a research report, the original obscurities and misdirections may already be smoothed over so as to offer a logical account that will fit smoothly into a projected official history. Still, even the significance of the findings in a research document will require time to develop. Some documents will disappear from history, others will be cited for some time as a basis for later experiment and theorizing, but all will disappear as working documents in a relatively short span of time. Controversy is essential in settling the claims of research reports for inclusion in the basis for further current experimentation. At a given point in time, then, a set of papers is presented for possible inclusion in the basis of science for further experimentation and theorizing. None of these claims can be guaranteed by methodology, but some of them will become scientific facts as a result of a process of legitimation in which scientific peers will negotiate their worth.[1] Foundational views in epistemology have been wrong, therefore, in locating scientific fact in the methodologically certified observations of individual scientists. Sociological views have been wrong where they have suggested relativism of fact. Unlike other social institutions in which negotiation produces settled opinions, negotiation in science must confront the constant production of new scientific data text. Because of this, the distortion caused by bias and authority at any point in time should ultimately encounter disagreement with new text, and be resolved. At a given moment, new bias and authority may be influencing the perception of fact, but older bias and authority will be leaching out. Because of the constancy of the problem, proof of progress is difficult, but progress in correcting distortion does constantly occur. Over time, some of the claims originally proposed will survive this process to become scientific fact. These facts will be objective in a perfectly coherent sense, and while they will represent reproducible features of the world, they will not be realistic in the philosophical sense of providing a totally accurate description of the world. The view presented here thus grounds science in fact, and grounds it dialectically in the sense that the production of fact depends on the use of instruments embodying prior theory, as well as a process of negotiation in which theoretical background of all kinds can be used to assess the significance of data proposals. It is time to examine this process more closely.

A clear and well-argued version of this approach is to be found in Part II of Ravetz's *Scientific Knowledge and Its Social Problems.*[2] Ravetz's book presents the first coherent account of how scientific activity, which in its local form is always subjective and fallible, results in objective scientific fact as a result of a two-stage social legitimation process. The fleeting objects of laboratory perception, the tenuous marks of the data, become intellectual constructs with associated scientific objectivity as a social consequence of testing and (possibly) intellectual transformation. In chapter 3 of Part II Ravetz makes the very important point that scientific research is craftsman's work. The individual experimental scientist is *not* a machine, noting pointer readings, counting blips, but rather he or she brings highly developed craft skills to the relevant tasks. Data, and hence scientific information, are dependent on the sensitivity of the investigator, his or her feel for the right workings of the apparatus, the correct choice of the right tools and techniques for investigating a problem, and so forth. The section on pitfalls indicates how this means that the foundations of science in the individual experiment are not certain, that is, that there can be no guarantee for the individual researcher that he or she has avoided error. Chapter 4 is concerned to differentiate the craft work of science from other forms of craft work. In particular, it differentiates the craft work of science from other kinds of craft work by pointing out that the objects of the craft are relatively abstract and intellectual. Many scientific objects, such as pure gases and frictionless surfaces, are only related to objects in the real world, or are partly realized in the real world, without being observable *as objects* in the real world. Physics, which is often thought to be the most advanced science, has had the greatest success in this process of abstraction. Experiments in physics are usually performed on "artificially" prepared objects that do not occur "naturally" in the world in the sense that effort must be applied to isolate and prepare them, and special instruments must be invoked to interact with them if knowledge is ultimately to result. Thus far, we have looked at experimental activity as a specialized form of craft work performed on special scientific objects.

Ravetz also argues that reasoning in science, particularly at the growth points in science, is craft reasoning, a subtle blend of inductive, probabilistic, deductive, and analogical reasoning. This must be so at growth points because the interaction of data and theory is indeterminate. One does not know if the data are robust (i.e., roughly constant under slight perturbations of the experimental conditions), what the relevant data are, exactly, and what is evidence for what. Science assumes the familiar routine as these issues get settled, so that routine work can be made accessible to normal scientists—what Ravetz calls the "competence of mediocrity." On this view, superior scientists have possessed this aesthetic feel for their craft, and the unique social structure of science enables these tentative and difficult original lines of reasoning to become hardened into determinate research strategies accessible to development by normal scientists. The craft researcher has therefore at least some of the tenuous skills we associate with an artist. Ravetz offers in this connection a very interesting critique of Kuhn. He argues that many puzzles in the craft context of science cannot be anticipated as laid out in advance, but arise *in* experimentation or *in* thinking about unexpected data, and can be articulated as problems even though they can arise only in the craft "perception" of an encountered difficulty, rather than in a deduction from background paradigms. Scientific problems are more varied and less to be anticipated than Kuhn's analysis allows. The good experimentalist sees himself or herself as interacting with a difficult, contrary, changing "live object." This craft activity, with its "magical" aspect, *is* the appeal of science in the modern world for experimental practitioners, a fact that slips past philosophies of science based merely on scientific discourse, and it is the motor of scientific progress, a fact not sufficiently noted in other philosophies of science.

In chapter 5, Ravetz argues that the craft nature of science precludes a uniform methodology that suggests that scientific reasoning can be done in a semicomputerized fashion. It is possible that the view that scientific reasoning could be semicomputerized arises from taking physics (or mathematics or both) as the paradigm of science, another assumption binding the positivists, Popper, Kuhn, Lakatos, and even, to some extent, Ravetz himself. Physics tends to develop discourse about populations of physics objects that are theoretically equivalent. For example, physicists may talk about electrons, photons, or whatever, classes that are very large and whose members are theoretically homogeneous with respect to their properties. Thus if a (random) sample of electrons or photons is obtained, and various properties are determined for this sample, it is methodologically straightforward to generalize about the entire population. Since samples that are theoretically equivalent can be produced in many laboratories, this allows a division of labor in which "normal" science can easily contribute to the entire enterprise of physics. In the human sciences, and even in biology, most of the background assumptions tend to drop away. Except for identical twins, most people have different genotypes—hence different behavioral potentialities, and the addition of differential learning means that no two human beings are, in psychological terms, completely identical. The upshot of this is that the whole semicomputerized methodological possibilities for physics no longer seem relevant. Extrapolations from sample to population are tenuous, and the repeatability of experimental results is not foreseen with any confidence. Those branches of these sciences (behaviorism, sociological statistics, and so on) aping physics have not led us close to answers to the kinds of questions we would like to have answered, and except for these branches, the division of labor shown in physics doesn't arise. Fields are dominated by individuals from the complex welter of data about inhomogeneous classes that confronts them. This should make it clear that there is no obvious warrant for supposing that various sciences will show methodological regularities in any fine-grained analysis of their structure, even if such abstract features as logical coherence are sought everywhere.

Chapter 6 of Ravetz's book introduces the social processes that convert the fallible individual research report (or rather, collections of such reports) into scientific *knowledge*. This process has essentially two stages. The first stage is the journal referee system, which controls the possibility of publishing research reports, and the second stage is the practice of citing earlier work in a field when writing individual research reports. Ravetz reports this as a descriptive sociological observation. He doesn't argue that this system is the best way to enforce quality, and he doesn't pretend that citable and valuable work can't be lost in the workings of this system. It is simply that this has been the method of creating and authenticating scientific facts throughout the period during which science has been advancing at a suitable rate, and therefore it seems to Ravetz *a fortiori* to have been working pretty well. Indeed the philosophical objections that can be brought against the difficulties and hazards of the referee process and use of citation point to a negative bias through a conservative selection process, which at least has the virtue of tending to ensure that the facts that finally appear are worth having. A fact that remains must be relatively invariant under various methods of investigation, and it must continue to be found as new instruments are developed to gather data. It is interesting to consider whether this two-stage process has not in fact worked well historically, and whether it hasn't been a good compromise between the twin dangers for the evidential base of having too much information (including too many shoddy research reports) or too little information (even if less shoddy throughout) to fuel current research.

Ravetz's scientific realism makes facts, the basis of science, intellectual constructs that have survived this two-part legitimation process. The gap between this and any observationally foundational philosophical view is enormous. In Ravetz's view, it doesn't matter whether individual observations are theory laden or not; *they* are not the foundation of science. This is also why the teaching of science doesn't proceed inductively from "foundations" in individual experiments, and why scientists are not much disturbed by philosophical attacks on the foundations of some branch of their subject. The hard data of science are at the level of legitimated fact, not at the level of philosophical epistemology. Or to put this another way, what is a fact for the individual scientist (e.g., preliminary data and a hunch) is not (at least not yet) a *scientific* fact. It should be obvious that the scientific facts in this view are something like superfacts, that is, after having gone through the two stages of legitimation they are, in some intuitive sense, more likely to reveal reality, or to describe features of the world, than are the deviant observations of individual scientists. Ravetz's realism thus suggests an interesting foundational view for science of a kind that has not been explored in philosophical epistemology.

A practical problem is that it can be shown that important work may not be noticed and that minor work may be thought important, so that the sense of this judgment "more likely to reveal reality" requires discussion. Science is dominated at times and in places by "schools" of thought that are later seen to have been working in a direction that gradually gave out. Thus the realism here seems to contain seeds of relativism. But the historical examples show that while there are outstanding examples of good scientific work whose acceptance into the scientific canon was delayed by bias, we obviously don't have examples of truths that were never accepted into the scientific canon. There are many scientists, of course, who feel that their work isn't being accepted fast enough, but motivation in science has remained high, suggesting that the working scientist normally sees the legitimation process as an acceptable hindrance or delay, rather than as a barrier to truth. Work of citable quality will be, in the common view, published somewhere, and then the author will have a chance to win recognition for it. It should also be noted that the publication of journals in science (and in academia generally) encourages debate. Without being "impartial," editors will knowingly publish controversy and exchange in at least some small quantity—a factor that tends not to be so obvious in some religious, art, and political journals. All in all, then, Ravetz has made a good case that as a practical instrument traditional scientific legitimation has functioned conservatively on the whole, but superbly.

There remains the philosophical skeptic. It is logically possible on the basis of Ravetz's realism that some important fact might be noticed by some individual who is incapable of communicating it to other scientists because of their conceptual blindness. But can this individual be certain that he or she is right, that he or she possesses knowledge where others have merely opinion? The longing for methodology continues to reassert itself. Seen against the background of old movies in which a hero scientist is trying unsuccessfully to prevent a loathsome epidemic among ignorant villagers by some, to their minds, devilish scientific instrument, this may seem plausible. It is perhaps less plausible but still possible where a scientist is trying to convince other scientists about the worth of selected data. The plausibility it has here is considerably diminished by the problem of knowledge at growth points that Ravetz emphasizes. My single observation that the sun is shining, or that some other state of affairs obtains, may be knowledge in an established context where the language used to describe matters is fairly settled and there is wide agreement about correct observation. But at growth points the evidence is typically so contradictory and complicated that those working in an area are well aware of the existence of problems and pitfalls and know that they need to strengthen partial views by more experimentation and by argument against other partial views.

We shall now examine the stages of the legitimation process somewhat more carefully, beginning with the dubiety existing at the level of experimentation. For this purpose, a detailed case study by Holton will be employed.[3] The point of this case study for our purposes will be the enormous complexity of craft manipulation required to get reliable results, and the sheer repetition of results until their stability was assured. Philosophers and sociologists have frequently simply overlooked the amount of effort required to produce the sort of classic experimental result grounding a scientific fact. Holton's case study concerns a dispute that occurred between the physicists Ehrenhaft and Millikan about the value of the smallest electric charge, the negative charge of a single electron. This dispute began in 1910, and its importance is related to the way in which the value of the negative charge of the electron is related to the calculation of other fundamental physical constants. Millikan and Ehrenhaft were engaged in similar lines of experimentation, but Millikan found a stable smallest value for the electron, while Ehrenhaft found subelectron values, negative charges of a half, a hundredth, and even smaller fractions of the electron's charge. This dispute lasted for over twenty years, although with historical hindsight it is easy to see that Millikan was right and Ehrenhaft wrong. At least it is easy to see this provided that modern postulated subelectronic charges do not become a reality, although even in this case it seems relatively obvious that these charges should not have played a role in Ehrenhaft's experiments. The dispute is made more interesting by the fact that Millikan observed subelectronic charges from time to time, but dismissed them as artifacts caused by experimental error. From the contemporary viewpoint, Ehrenhaft was also the more distinguished physicist. Of further interest is the fact that the general emergence of the stability of Millikan's experiments permitted scientists to "see" individual electrons, a fact cutting against the view of phenomenological physics, which held that atoms and smaller particles receive ontological significance only as organizing principles for direct observations in experimental settings. In short, this dispute has the attributes of emerging significance that have been argued for in our historical discussion.

Millikan began his research with the conviction that there was a unit charge, and that the experimental means for determining it were fairly obvious. Before Millikan's work, those proceeding on a similar assumption had only been able to calculate the average charge on a population of electrons. Millikan began along the same track, with the idea of simply making small changes in existing procedures in order to improve accuracy. Water droplets were produced in an expansion cloud chamber in the standard procedure, some then falling at the gravitational rate, but others (charged) falling more rapidly in an electric field imposed on the droplets. A number of simplifying assumptions were required to interpret this experiment, such as the assumption that the droplets were similar or that ionization did not substantially affect the size of the droplets. Millikan and a collaborator began by ionizing a moist gas prior to cloud formation in a new way, and in so doing they substantially improved previous results. This newer value was of help to Rutherford in another connection. In discussing Millikan's work, Rutherford suggested that it could be further improved, and pointed out in particular that evaporation in the experiment was probably underestimated, causing the calculated value of the charge to be too small.

To study the effect of evaporation, Millikan used a much stronger battery to create an electric field that could hold the singly charged layer of droplets steady against the influence of the gravitational field. The result of this stronger field was to destroy the observable cloud, including the layer of presumably singly charged particles. What happened, as it turned out, was that the droplets were actually variously charged, and the field caused them instantly to separate, hence the disappearance of the cloud. With this exposure of a false supposition, a decade (note the time period) of cloud watching as a means to observing the value of the charge of the electron came to an end. Then Millikan noticed that although the cloud was dispersed, some few droplets remained in place, stationary in this strong field, presumably just the droplets that were singly charged. Attention now shifted to watching singly charged droplets held stationary in an electric field, instead of a falling cloud of droplets. Millikan also noticed that some droplets, apparently catching a new charge, would suddenly begin to move decisively. Observing these suddenly moving droplets, Millikan calculated that charges seemed to come in integral multiples of the basic charge, confirming the conviction of unit charge.

Millikan's first major paper on the new method was presented in 1910, and included a critique of other methods of measuring the unit charge, including Ehrenhaft's. In his paper, Millikan divides his observations into those that were "best," those that were "good," those that were "fair," and those that he discarded. The craft nature of observation is manifest. Millikan relied on his feel for the correct working of his apparatus, and was also constrained by his anticipation that good observations should cluster around the correct value. Better observations were therefore weighed more heavily by Millikan in working up the data to produce a calculated mean value for the unit charge. Ehrenhaft, reacting to Millikan's critique, began a series of new experiments utilizing a quite different technique from Millikan's in which he reported fractional values of the electron charge. Further, he subjected Millikan's paper to a countercritique, showing that a differing mathematical workup of Millikan's data actually supported the doctrine that there were subelectronic charges. Clearly, where Millikan saw divergencies in his data as clustering around an assumed correct value, Ehrenhaft saw Millikan's data, like his own, as a dispersion of values, indicating fractional charges.

At various times, Millikan tried droplets other than water droplets, but the breakthrough to his famous experimental determination of the unit charge of the electron came when he tried utilizing oil droplets rather than water droplets. This minimized the uncertainty of various theoretical assumptions, such as the assumption that evaporation was not affecting the experiment, and the stability of oil droplets allowed Millikan to boast that he could observe a fixed number of charges on such a droplet as long as he liked. The previous dispersion of values literally disappeared, and the number of observations that had to be considered poor, or discarded, dropped enormously. In the meantime, equipment components were improved, a better optical and timepiece, as well as better temperature, pressure, and electric field control. Although improvements were made for years, the first run of oil drop experiments dramatically reduced the probable error in the calculated mean of the unit charge, and forecast the eventual resolution of the controversy over its existence. After five years of pursuing some form of droplet experiment, Millikan was now ready for a long series of runs on the oil drop experiment to clinch the view that it established the unit charge beyond question. In one six-month period (note the time interval) from October 1911 to April 1912, Millikan ran 140 experiments in which individual oil droplets were observed as they picked up charges. A paper published in 1913 on this series (and on other series) that singled out 58 drops as the data base for a mean value produced a value that Millikan accepted for twelve years, despite constantly improving technique. Many of the drops studied were not included in the data base because of suspected equipment or observational failure. These judgments were not included in the 1913 paper, as they had been in the paper of 1910. Millikan's self-doubts were obviously a thing of the past. The better technique allowed him to more sharply distinguish good results from bad results, and his confidence also measurably increased as it became clear that the unit charge could be more successfully integrated into the body of physical theory than Ehrenhaft's fractional charges, leaving as the only open question its precise determination.

This episode will stand here for many that might be taken from the history of science. An early line of experimentation is almost accidentally selected as promising, and then refined in a series of trials and errors, as well as by the consequences of at least two sharp insights, the method of using a strong enough field to hold the drop steady and the switch from water to oil droplets. There is a convergence from vague perception to a reliable technique, and then a sustained series of concentrated observations clinching a basic value, at least in the theoretical context of the times. The constant craft nature of the scheme is shown in the fact that Millikan's experience gradually produced clear data ("beautiful," as it is described in his notebooks) although the equipment cannot produce clear data for the uninitiated, for example, contemporary beginning students of physics. Controversy with Ehrenhaft and the constructive criticism of Rutherford gradually set the direction in which experiment took place. Original dubiety about the calculated value, coupled with exposure of a wide range of data, gradually became transformed into near certainty about the value, coupled with exposure only of the data relevant to that calculation. In time, this method of calculating the unit charge produces the value that is quoted in other work as a fact on the basis of which other calculations can proceed. These are the features typical of the development of a fixed experimental datum that our abstract discussion has led us to expect.

We can now turn to the conversion of subjective judgment into objective fact over the life of citation of some proposed finding. The initial filter discussed by Ravetz is the process of refereeing before publication. Success in this process is recognition of the fact that in the opinion of one or more peers, work is of more than minimum quality, making it a possible object of scientific controversy and scrutiny. What this process lacks in objectivity is shown *informally* by the fact that a paper may be rejected by one journal and then accepted by another journal. Within disciplines and research areas, an informal ranking of the quality of journals may cause a tendency for articles to be submitted to journals of decreasing perceived quality, so that final acceptance is a partial measure of the strength of what would be consensus about the paper's quality, but this has been little studied. We are interested here in the change of modality that occurs when work becomes more widely cited, and the author or authors discover that recognition for their work is increasing. Fleck had noticed in 1935 that journal articles may contain modalities such as "I have tried to prove . . . or "It appears possible that . . ." or "It cannot be conclusively established . . ." that are consonant with the stage of controversy over the significance of contents.[4] As significance becomes settled in science, and the resultant facts become fixed points in review articles and later work (and ultimately in textbooks), the modalities are stripped away until only assertion remains. Fleck, studying the relationship of the Wassermann reaction to the concept of syphilis as a disease, noted that the correlation of positive Wassermann test results to confirmed cases of the disease rose as the reagents and timing of the test were manipulated until the test could generally be seen as reliable.[5] The literature involved in the serodiagnosis of syphilis by means of modifying and simplifying the Wassermann test reached, on Fleck's estimate, ten thousand journal articles.[6] In this case, the development of technique and the phase of controversy before consensus about fact are enormously complicated.

This point can be developed in greater detail using an example from an analysis of scientific work in biomedical research at the Salk Institute by Latour and Woolgar.[7] After a discussion of how controversy shifts the status of scientific statements between various layers of a typology ranging from mere conjecture to assertion of fact, Latour and Woolgar offer an analysis of the construction of one scientific fact: the discovery of the structure of a peptide whose function in the endocrinology of certain mammals is the release of thyrotrophin. The potential value of this fact lies in its role in helping to explain how the human brain may control human behavior by utilizing such releasing factors. This discovery was made principally as the result of research undertaken in two rival laboratories between 1962 and 1969. From 1969 to 1971, this discovery was frequently cited by other researchers in their papers. A marked decline in citation from 1971 to 1975 occurred as the suggested structure became a fact that could be taken for granted, and by 1975 textbooks reported this structure as a fact that had been established by the two laboratories in question. Ravetz's account of citation requires some refinement to allow for this pattern of increasing and then decreasing citation followed by textbook representation, which may well be a general pattern for scientific literature, especially successful literature.

Biomedical research in the relevant area consists of attempting to match the action of natural substances to that of synthetic counterparts. A natural substance may have a known effect on an animal, or on the cells, muscles, or whatever, of such a creature. This substance can be split into subcomponents and the effect of the subcomponents on the animal, cell, muscle, or whatever compared to that of the original substance. In this way, which is methodologically crude in its logical structure but tremendously sophisticated in its instrumental realization, researchers attempt to isolate and purify the entity in the natural substance responsible for the effect of interest. The purified entity can be analyzed partly through the use of various techniques, but the exact structure of the entity cannot, in general, be determined by analysis. To find the exact structure, researchers attempt to synthesize the entity from pure amino acids (and other building blocks) by techniques that allow them to understand what they have synthesized. This process of synthesis is guided by the clues provided from analysis. When a substance is synthesized that matches the effect of the purified entity from the natural substance, the structure of this purified entity is then thought to be understood. This process can be nearly dialectical when analysis restricts synthetic possibilities, whose exploration can restrict analytic possibilities, and so on. Further, after identification, slightly divergent synthetic forms can be explored for their potential in therapeutic medicine.

From 1962 to 1966, most of the work on the substance of Latour and Woolgar's example was based on the expectation that it was a peptide. Early research showed that the action of the substance was destroyed partially or totally by certain enzymes and by heating with hydrochloric acid. These crude tests for peptidic nature resulted in the following language:

> In this note we show arguments in favour of the peptidic nature of these substances. . . .[8]

As the language shows, this evidence was not conclusive. In the next few years, further work, which failed to show an increase in the ratio of amino acids as the extract was purified and failed to destroy the effect of the substance with more sophisticated enzyme tests, led to the tentative conclusion by one of the laboratories that the substance was *not* a peptide:

> We have been led to question the long held hypothesis that [the substance in question is] of peptidic nature.[9]

The other laboratory, now entering the field, also confirmed the new hypothesis that the substance was largely nonpeptidic, but it was puzzled by its own discovery of the presence of amino acids in a percentage suggesting peptidic nature:

> The results are consistent with a hypothesis that [the substance] is not a simple polypeptide as has been thought previously, but nevertheless our evidence indicated that 3 amino acids are present in this molecule.[10]

In reaching this conclusion, the second laboratory was clearly influenced by the work that had been done in the first laboratory. By 1968, a number of new techniques imported from other fields and a massive amount of brain extract led to new results and a realization that the problem was enormously complex:

> Our efforts at characterizing the chemical structure of [this substance] have led us to the conclusion that we are dealing with a rather difficult problem for which classical methodology is turning out to be of only limited significance.[11]

In other words, the particular nature of the substance precluded the use of normal techniques, such as the use of gas chromatography, for which the substance was not volatile enough. Special techniques needed to be developed. The new techniques came from chemistry, and they revealed that the rejection of the peptidic nature in 1966 had been a mistake. The breakthrough came from the synthetic direction. Possible combinations of the amino acids found in the new analytic assays were tried and dramatically lessened the number of possibilities. Finally, a better way of analyzing the purified natural substance was discovered, and the possibilities were reduced to one. The purified natural substance was no longer said to be similar to or like the synthetic compound: rather, it was said to be identical with the synthetic compound, and this fact then became a given in research and a commonplace in textbooks. Both laboratories produced histories in which they were primarily responsible for the discovery, but they are now usually given joint credit.

This example shows quite clearly the shift in modalities as the relevant fact came into view, and it indicates that the significance of the research in 1966 was not apparent until better experimental techniques were available in 1969. A curious and not uncommon inversion in significance had taken place because of a changing background of accepted possibilities. Evidence toward a peptidic nature for the substance first became available, then evidence pointing away from this conjecture, and finally evidence for a peptidic nature as the evidence was measured against an increasing array of theoretical possibilities and experimental techniques imported from other fields and adapted to the specific nature of the problems in this research area. Curiously, after the structure was discovered, an enzyme was discovered that might have set the research of 1966 on the right track. Therefore, the evidence takes on an even further shifting quality when it is set against possible scientific backgrounds, and its historicity is confirmed by the inclusion of such possible contingencies in its evaluation. The discussion and examples of this section should be sufficient to indicate why it is more revealing to consider scientific facts as constructions than as mere careful observations, and why this shift in descriptive terminology does not mean a concession to any easy relativism.

Data Domains

Biological species adapt to environmental niches, but adaptation may be a complex phenomenon. Species may fit into neighboring or new niches successfully when such an opportunity exists because the niches are open, or because they are filled with other species that can be driven out. Species may also alter the niches to which they are adapting by changing them physically, or by altering the total ecological balance so that the relevant niche changes in terms of food possibilities or the nature of existing predators. Change need not occur solely on the side of the species. Environmental changes not attributable to the species, or attributable only partly to species activities, may produce a change in the nature of the adapting species, if only in the distribution of variety within the adapting species. Because of this dialectical interplay of species and environment, it is possible to argue that any sharp distinction between species and environment is a conceptual illusion. Such illusions, however, may be necessary at least on a temporary basis for understanding. To some extent, the interaction of species and environment is simplified by the perceptual apparatus of the species, which simplifies the total range of possible environmental changes into a smaller set of signals to variations in which the individual members of the species have the capacity to respond. This fact allows the internal workings of the members of the species to be studied against limited environmental changes. Such studies allow insight into the means by which individuals can respond to environmental change and the limits of such means. Although theory and fact are locked in dialectical interaction on our view, a similar conceptual framework can be used to freeze the data and study the response of theory to fixed and slightly varying data. Similarly to the biological case, there are good methodological grounds for fixing the data and studying the means by which theory can adapt, rather than the reverse. The relative simplicity of the organism or theory with respect to its environment allows one to project a comprehensible possible range of adaptive tactics against a relatively fixed environment, whereas the total range of possible environmental change must remain largely unknown.

If the sensory apparatus of organisms simplifies environmental change and reduces it to a set of signals to which the organism can respond, it is clear that successful species will have sensory apparatus that is, in some sense, matched to those features of the environment that are most indicative of the biologically relevant aspects of its environmental niche. Success will depend on the availability of a suitable range of responses to change, but it also depends heavily on a suitably discriminating sensory apparatus, one attuned to the most important environmental indicators. Theories adapt to fact in similar ways, refining the facts so as to yield a conceptually clear version of their significance, and then projecting a range of possible facts attached to possible new observational locations. At any given time, the set of possible facts will be enormous, and a theory will typically reduce this set by discriminating between relevant facts or relevant features of facts, or both, and those facts or features it will ignore. Given what we have said about the significance of fact, some theories will take a losing gamble. They will adapt to facts or features of facts that do not stabilize as the environment is explored and perception becomes more refined and sensitive. In the long run, the projections of such theories will prove unreliable, or less reliable than those of theories that happen to fasten on the facts that do stabilize and are significant in revealing environmental change. Our historical point that facts emerge from a social process in science reveals why Popperian falsifiability is such a misleading notion. To falsify a theory, the significance and stability of a fact must be given with its putative isolation. A specific fact may be lethal to a very specific theory making an extremely specific prediction, when it is isolated, but the more usual case is that the isolation of a putative fact will not destroy a theory, which is best conceived as a bundle of specific interpretations of a set range of possible facts, just as a species is a bundle of specific genotypes, but it will tend merely to change the distribution of its interpretations. The theory may constrict the niche in which it is viable, but to drive it to extinction will require a great many adverse facts whose permanent significance is regarded as settled, as well as perhaps suitable competition or predation.

The total possible set of environmental niches for species can't be described. What we can do is match existing species to the niches in which they have proved viable and project possible niches in which they might also succeed. When this is done, it is clear that the available niches usually have shifted over time, but also clear that some available niches are not filled until suitable organisms develop. If we wish to define data domains as factual niches to which theories adapt, by analogy to the environmental niches to which species adapt, we have to deal with the problems of defining environmental niches that are shifted to the new area of consideration. The problem of individuating niches has a number of serious aspects. The individuation of niches is constrained in nature by the biological possibilities because new organisms adapting to such niches must somehow result from the possible reproduction (allowing for mutations and meiotic scrambling) of existing forms of life. In the scientific case, discriminations of mutations from new forms becomes much more a matter of convention. In both cases, we have the problem of separating true niches, into which viable species can adapt, from false niches, where species will fail because their sensory apparatus and repertoire of responses are insufficient to continue life.

Man's environmental niche was originally constrained by his sensory apparatus plus the discriminations in perception that could be coded into ordinary language. This niche has been considerably broadened by science. The instruments of science can best be seen as refining and extending human sensory apparatus, and scientific languages as refining and extending the discriminations that can be coded into ordinary language. As a result of these extensions, more people can remain alive, and can remain alive under more varied circumstances, than would be possible without science. Specialized theory, like any specialized conceptual apparatus, will permit relevant discriminations to be made and acted on in individual data domains, but the scope of theory will be related to the extent of such a domain. In technology and special skills such as cookie baking or horse breeding, an activity will be extended and refined. In science, it is our pure sensory apparatus that is extended and refined. Telescopes and microscopes extend human vision, and not a particular activity, and they provide new objects that science can see, describe, and theorize about. The individuation of data domains thus stems ultimately from the human sensory apparatus and its instrumental refinement and augmentation in science. Objects of ordinary perception become separable into objects of extraordinary perception, but extraordinary perception, even when it seems to be dealing with an autonomous world, receives its significance from significance in ordinary perception. Modern sophisticated measurements of length, for example, have humble origins in such beginnings of instrumentation as are represented in this sixteenth-century instruction:

> To find the length of a measuring rod the right way and as it is common in the craft . . . Take 16 men, short men and tall ones as they leave church and let each of them put one shoe after the other and the length thus obtained shall be a just and common measuring rod to survey the land with.[12]

Because data domains must be individuated for the adaptation of theory, and because they are constructed through instruments whose most sophisticated operations are ultimately read by the human perceptual apparatus, some kernel of anthropomorphism remains in even the most advanced science, no matter how little subjectivity remains. Further, just as sight may yield vague, confusing, or ambiguous data, and sight and hearing (or any other pair of sensory modalities) may conflict, these extended data domains cannot be expected to be free of perceptual problems and the attendant process of eliminating those theoretical conjectures that misread the perceptual field.

The objectivity of certain features of the world is given in science as it is in ordinary perception by the repeatability and robustness of perception. That my house is on the east side of my street is supported by its always being there whenever anyone looks. In the extended and augmented perception of science, repeatability and robustness cannot be taken for granted. Instruments must be developed that will allow repeated observations to be made and will locate the same features on repetition, at least for trained observers. Objective observation of fact is not in question until the level of fact described as a result of the social process of negotiation in the last section is reached, and until instruments are available that permit the fact in question to be ascertained by scientists with divergent interpretation of theory. Here we will stress the role of instrumentation in fixing the limits of data domains and in locating the facts within them.

People with different interpretations of situations may well see differently, and it has not proved philosophically fruitful to attempt to find a level of seeing at which everyone is said to see the same thing, so that differences are explained as differences of interpretation of this basic level.[13] Now let us consider an everyday example. Two people who wish to move a desk through a doorway disagree about whether the desk will fit through the opening. The person who believes that the desk will not fit may see the desk as being larger than it appears to the person who believes it will fit through the opening. Our interpretations of situations, our beliefs about situations, and our theories about data may influence the perception of situations and data. In fact, since it will do so, the absolutely neutral observation of fact required in many forms of empiricism and rationalism is an illusion and cannot be proven to exist. Empiricism and rationalism could not prove its existence, but depended on a transcendental argument that such observation was necessary to ground the objectivity of science. These arguments we have laid aside. It is interesting to notice what an instrument accomplishes. An instrument breaks the line of influence from interpretation to observation, or from theory to fact. Both persons interested in moving the desk may locate a ruler of some kind, and then be in a position to agree that the measurement of the desk is sufficiently larger or smaller than the door opening to settle the question of whether the desk will fit without trial and error. The measurement may result in the discovery of the fact that the desk is narrower than the door, or vice versa. Incidentally, given the hazards of measuring, a sufficiently similar measurement for the desk and the door may not settle the argument, and even the settled argument may be mistaken because of some other fact, for instance, that the door can't be opened widely enough to expose the whole doorway to the desk because of some feature of building design. The use of the ruler also doesn't reveal a brute fact, but rather a shared interpretation or belief, since the ruler can establish the relevant facts only if it is agreed that the ruler is of the same length in both relevant locations, and so forth. Naturally, this agreement may be tacit, and not consciously arrived at. What we have here is a situation in which opposed viewpoints must yield to the fact that the desk is about twenty-eight inches wide (let us say), and the doorway thirty inches wide. The world is still interpreted, but the individual interpretations have been modified.

Scientific instruments function like the ruler in the example, at least when the significance of their use is widely accepted. Instruments function to break off the influence of assumption on personal observation. If they did not exist, the fact of the influence of theory on perception might mean that shared data would be impossible. Where they do exist, a level of objective scientific fact of the kind analyzed in the last section is more likely to be achieved. The instrument is usually a machine. Properly used, it stays the same from experiment to experiment, and its reduction of data complexity to a common signal, like a number expressing a reading, tends to neutralize what may be a highly variable scientific temperament from experiment to experiment. In physics, where instruments are plentiful, domains of scientific fact are numerous and in many cases clearly individuated. In sociology, where instruments are not plentiful, theory tends to penetrate through to observation, so that two sociologists may see the same limited situation in quite divergent terms, and instrumental reduction of the divergence may not be possible.

Another way of marking the influence of instruments on the objectivity of data, breaking the influence of theoretical expectation on observation, is the following. As we have repeatedly observed, the data gathered at the growing edge of science are confused and uncertain. It is exactly in this area that theory can be seen to guide perception. Bacteria, investigated by the early microscopes, were at the limits of resolution, and could be seen as having almost any shape. Saturn and Mars, investigated by the early telescope, were of ambiguous configuration, and could be interpreted by different investigators as having divergent features. The existence of new particles in elementary particle physics can depend on whether a double peak in a graph is deeply enough separated on a histogram. At the limits of instrumentation at any given time, who can blame scientists for finding confirmation of those views to which they have committed their research, even if this confirmation is muted in the research report?[14] The improvement of instrumentation can remove this ambiguity. Where expectation had caused divergent perception, a common construing of the data may take its place.

The ruler in our desk-moving example can be used in connection with many different problems. Similarly, many general scientific instruments—the telescope, microscope, spectrometer, voltmeter, chronometer—are used in a variety of scientific investigations, and they presumably work the same way in each of their applications. This helps to circumvent a problem about falsifiability that has been proposed by many authors. A theory will, in many cases, predict an observation only in the context of auxiliary theories, at least some of which are involved with the instruments required to make the observation. If an observation incompatible with the theory is made, it can be argued that the blame could be assigned either to the theory or to the auxiliary theories. But auxiliary theories involved with instrumentation may also occur elsewhere in science, and the scientist is not in a position to argue that they are false only in his or her area of research, provided that the instruments are working well elsewhere. This fact constrains falsification in the direction of the primary theory and away from the auxiliaries, and blunts considerably the mere logical point that the falsity of the conclusion to a valid line of reasoning may be the result of the falsity of any of the premises. That instruments are used by different scientists in the same research area and are also frequently used in different areas of research means that the facts they uncover, and that they produce such and such data in such and such circumstances, is not under the control of the individual scientist. No matter what facts are agreed on, the choice of a theoretical interpretation of those facts from available alternatives is a matter of personal decision. A scientist may choose to evaluate new instrumental findings as artifacts of experimental design, and as not requiring theoretical interpretation, but this involves the gamble that these findings will not survive the legitimation process and become scientific facts. One can perhaps more gracefully change theoretical stance in the light of new facts than one can admit to having looked at solid findings and set them aside as errors.

The independence of theory from fact that instruments help to create is reflected in the history of instrumentation. Scientific purposes for which the instrument is to be used need not be involved in any detail in refinement and improvement of the instrument. Refining instruments is a craft process and is not necessarily led by theoretical reflection. Aside from the obvious fact that the light telescope was to magnify distant images, improvements in the smoothness of surfaces of lenses and mirrors and in the means of mounting and moving the telescope to ensure solidity could be carried out quite independently of the theoretical curvature of light to be interpreted by the instrument. To a certain extent, the working of the instrument can be explained in ordinary language, and the working of the instrument is not dependent on the data augmentation and theoretical advances its use helps to bring about, along with the specialized languages required to express new data and theory. The telescope augments human visual perception. Without long-exposure photographic complementarity, the telescope allows one to see what could be seen from a different and closer position with the naked eye, especially when the constant distortions and visual artifacts of the telescope system are filtered out through familiarity with its use. Different interpretations of the resulting observations are possible, but the resulting observations cannot at first be influenced by full theoretical expectation. As is well known, the first observations of Saturn produced a variety of interpretations of its structure, a variety that was quickly narrowed down as a result of theoretical advance.[15] The early interpretations occurred in an earlier language, and were not yet in the full grip of theory. What is seen as a new kind of data, such as the data first revealed by a new instrument, must be described graphically, as like what is already known. As theory develops, description can become much more detailed, but expressed in the descriptive terms of the new theory, with a resulting loss of vividness.[16] As will become clear, instruments, data, theory, and language all tend to evolve into close consonance in many areas of science, but instruments play the essential role of initially separating data from theory so that a direction of accommodation can be plotted.

If the importance of instruments as stressed here is correct, why do histories of science and books on the philosophy of science pay so little attention to the details of instrumentation? The indexes of such books typically include an overwhelming number of entries to people and to ideas, but not to specific instruments. Many books are written with titles like "The Conceptual Development of Quantum Mechanics," or "The Conceptual Foundations of Quantum Mechanics," but very few with titles like "The History of Instrumentation Used in Quantum Mechanics." In most books of history or philosophy concerning science, experiments are described in only the most abstract terms, for example, "In this experiment, a beam of electrons was passed through a narrow slit and . . ." Some means of reconciling this fact with the importance of instrumentation stressed here is requisite.

Although instruments produce new data, some of which become legitimated scientific fact, we have already argued that the legitimated scientific fact is the foundation of scientific epistemology. Instruments, which in many cases are necessary to produce this level of fact, lie below this level once it is established. The data and legitimated fact guide the development of theory. Since there are points at which data become fact, these can be treated as fixed points for theoretical development. Of course, theory influences the development of instruments, but theory never becomes fully legitimated at quite so narrow a moment in time as fact. Instruments, like eyeglasses, are used to see things, but they need not be noticed (unless they malfunction or break) once what is seen through their use takes on independent existence. When the significance of a fact is known, the fact is important, but the details of its discovery no longer matter save as a source of possible heuristic advice for other scientists, or as part of the history of an area of science.

A more important omission of reference to the details of instrumentation may seem to be the relative lack of attention given this topic in journal articles. An instrument can be named or illustrated in an article, or a modification in technique noted. These sorts of references do occur. But the feel of an instrument and an explanation of its working cannot easily be accomplished verbally. Theory and data can be described, but one must learn how to operate an instrument. In most research areas, a common set of instruments for investigation is a given of research strategy, and new instruments are generally borrowed from known research areas. An understanding of the workings of these instruments and their limitations is part of the common knowledge of research groups, a fact insufficiently noted in Kuhn's discussion of paradigms and in other attempts to explain unity in scientific understanding. Any attempt to develop this for outsiders would run into the difficulties of explaining technique verbally, but there is no need to explain this to outsiders in journal articles, since they are not addressed to outsiders. Learning to understand instruments and their use is a staple of scientific education, and is accomplished appropriately in the laboratory practice of laboratory sessions run in conjunction with lectures on the relevant scientific topics.

In the view of history proposed here, instruments are important at the growing edge of science. Their primary significance is to contemporaries. They define the kind of data it is possible to obtain, and the limits on that data. When the data have been legitimated, they drop out of consideration. This is probably why instruments have not been noticed by philosophers as a real constraint on data text, and it helps to explain why reference to instruments does not occur where one might initially anticipate it. Our understanding of scientific history, however, cannot overlook this feature.

Data domains cannot be defined in terms of instrumentation alone. Some instruments are used to define different domains, possibly in conjunction with differing sets of related instruments. We have emphasized that new instruments can produce new data domains or be developed to explore new data domains, but we need to distinguish between the refinement of an old instrument and the development of a new instrument, no matter how blurred this line might be in practice. We can't have difference in domains distinguishing new instruments from refinements and also pretend that new instruments, rather than refinements, individuate new domains. We have been working with a basic analogy between theories and species, on the one hand, and data domains and environmental niches, on the other. The problems with discussing the relationship of species to niche have already been mentioned. An alteration in a species may change the niche it adapts to, and the discovery of a niche by placement of a few members of a species in it, perhaps because of an accidental translocation and subsequent impossibility of returning to the old niche, may transform the old species, possibly into a new one. It is hard to regard either side of such an interesting pair of concepts as fixed. In the case of theory, because of the same sort of historical considerations, it will be adapted to, and adapting to, a data domain whose nature and extent have some permanent features given by legitimated fact, and some newly discovered features given by a set of instruments regarded as in interaction with the set of objects the theory is concerned to talk about. A variant of a theory can suggest a new or an enlarged niche for itself. At the same time, a new niche may be encountered by instrumentation, and representatives of older theories may enter such a niche, but then undergo such drastic change that they transform into new theories. But the temporal sequence may not be so crude, and the interaction of data domain and theory may be extremely complex. As in the case of speciation, any change in theory (or in the distribution or nature of its subordinate interpretive variants) may cause a change in its related data domain, and vice versa. How extensive a change is required before we recognize a new domain or a new theory is a matter of conceptual convenience that has little to do with subtle variations in the real process being described. Just as wild changes in developmental forms, or forms for different times and locations, can be seen as variants of a species, because the niche being adapted to in some sense requires these forms, so wild change in theory can best be seen in many cases as adaptive response to a changing data domain. Two variants can be seen as different versions of the same theory, variants of two conflicting theories or totally distinct in terms of the data domain to which they are directed. Logic must simply view them as distinct and lose the process of adaptation over time.

The problems can be illustrated with the light telescope. Theories of the nature of the universe existed before the telescope and were based on naked eye observation. An understanding of how heavenly bodies are related, how they move, and so on is a good example of interpretation of a domain given to all human experience. Before Galileo's use of the early light telescope, his predecessors had used a few instruments, but only to support and refine naked eye positional observations of the heavenly bodies.[17] Galileo's use of the telescope clearly opened a new domain in which, for example, the phases of Venus and the existence of sunspots called for dramatically new kinds of theory. Galileo himself regarded the telescope as adding a new and superior sense to man's panoply of natural and common sense. The telescope thus marks a sharp break in knowledge. Even if the Copernican turn could be argued for in terms of theory applied to the old measurements, telescopic observation of new kinds of objects relentlessly produced totally new kinds of theories about heavenly objects. But new objects were discovered as the light telescope was refined and augmented with timed photography. Galaxies, star clusters, nebulae, and so forth, objects as new as the phases of Venus at the time of Galileo's discovery, were discerned gradually as the telescope was improved. Theory changed in response to these discoveries, but the immediate problem is whether the refinement of the telescope produced new data domains, without the telescope becoming a new instrument. Because the theory of the expanding universe, developed to deal with the discovery of galaxies and their observed motions, is quite incompatible with most previous theory, it would seem compelling to see new data domains here, and that could be justified by the inclusion of new instruments and techniques that were required to measure the red shift, and so on. The point of introducing the notion of data domains is independent of the details of resolving such puzzles. Without the notion of data domains and the associated notion of how instruments are essential to their exploration and to defining their extent, the history of the interaction of theory and data becomes a series of arbitrary adjustments in which the constraints felt by contemporaries are totally absent.

Any full account of instrumentation would require a discussion of the fact that instruments are developed that are quite general in nature, such as the telescope, microscope, and spectrometer, that can be adapted to research in many different areas, and that for the quite specific purpose of demonstrating some important fact, such as Atwood's machine for demonstrating Newton's second law.[18] These latter instruments may have a complicated evolution with respect to the data they are designed to obtain. A more general instrument will be developed along common lines, small advantages and refinements occurring in a manner that can usually be anticipated, but a manner subject to the problems of craft improvement. In the development of the laser, as reported by Collins, there were many laboratories attempting construction of a more powerful gas laser by increasing the operating pressure of the device.[19] The general usefulness of the device was recognized, and attempts to increase pressure without obtaining breakdown in operations were proceeding along uniform lines, with many laboratories constructing essentially the same design. Operation of the laser is uncertain, at best. Sometimes it will work satisfactorily, and other times not, without explanation, and a model built to the specifications of another will not perform in the same way. This is an interesting example of the problems of operating equipment, and Collins is interested to observe that competition between these groups resulted in less than ideal communication between them. What is interesting here is that communication was not perceived as needed in many cases because the general design was a given, and the various groups building such lasers were looking for variants that would result in better working through a trial-and-error or craft approach. For many such general instruments, one would expect different instruments to be constructed on different sites, all of which might be regarded as variants of the same instrument.

The search for an instrument to demonstrate a specific fact can be quite different. Collins also reports on a variety of attempts to construct a device to observe gravity waves in order to confirm the general theory of relativity.[20] Here a race is on to discover gravity waves, predicted by a theory, in a manner that will make their existence a legitimated scientific fact. In this area, at the time when Collins wrote, one person had claimed to obtain data showing gravity waves, but this did not produce a rush toward design of a similar apparatus. The data of the first scientist had only been partially communicated, but it was clear to many of his competitors that his data might not, or would not, produce legitimated fact. They therefore explored different kinds of apparatus in an effort to find convincing and repeatable data. Even those who were tempted to accept the original finding were not anxious to be second. In all of these efforts, there was a conscious intention not to produce the same apparatus in order to have an opportunity to receive recognition for first establishing this important fact. As we have seen in the case of Millikan and Ehrenhaft, two scientists expecting different data to be legitimated may quite consciously choose to investigate some given data domain with quite different instruments. These examples may be sufficient to establish that the history of instrumentation would be no passive narrative, and that the relationships of the instruments developed to the data domains they explore may be quite complex when viewed in detail.

The Microprocessing of Scientific Fact

Continuing the biological metaphor that has played an important role in this chapter, we may consider the case where the environment changes very little over a period of time, and the species filling a niche in this environment has sufficient genetic flexibility to adapt closely to its niche. Under these circumstances, the species and the environment will reach a close relationship of adaptation in which the species will attain a distribution of members that will remain stable over time. The sensory apparatus of the species will also show conformity to the major features of the environment requiring response for appropriate homeostasis to exist, and a delicate balance of properties on both sides can be achieved not unlike that between symbiotic partners. In the scientific case, there is a gradual coincidence of the properties of theory, data, and instrumentation where appropriate theoretic flexibility exists to match the texture of reality as revealed in the gradually legitimized data obtained from suitable instruments. This is hardly surprising, since it can be viewed as an extension of linguistic adaptation to prominent features of the surrounding world in ordinary life and language, although in this case the sharper split noticeable in science between theory and data is more difficult to discern. Of course this difference is precisely what is to be expected, since ordinary language vocabulary can be expected to have resulted from the settlement of many disputes and the accumulation of many observations about those features of the world of most importance to a wide class of human beings in the pursuit of their interests. In this section, we will consider in more detail this inherent drive toward coincidence of language and fact where only small perturbations of data and slight divergencies in theoretical outlook are in question. The full dialectic of accommodation of theory and fact will occupy us in the next section.

Our running engagement with some of the major philosophies of science receives an interesting update in this connection. What Kuhn regards as normal science and sees as the typical activity of the majority of scientists is closely related to the phenomenon to be discussed. In a great many areas of modern science at any given point in time, theory, data, and instrumentation will be in sufficient consonance that current work will be appropriately seen as an extension of previous work, fitting theory more closely to data, and data more closely to theory, with improved and refined instrumentation an important desideratum in the process. In this context, a few anomalies can be seen as of little immediate importance. The paradigmatic background is not necessarily a shared set of values so much as a shared data domain and style of language, the two having adapted sufficiently that fact and value, data and theory, have become inextricably woven in the activity of scientists. Science at a local level need not have progress as its goal, but achieving fit between language and fact. In medicine, for example, research programs may seek a test to define a disease and definitive treatment for the disease so defined. A fully successful small research program will degenerate and stagnate as it attains success, leaving a known cure as a residue. Investigators will then turn to other research problems. In physics, a similar phenomenon is to be observed. The laws of the inclined plane are settled, and textbooks may refer to fixed solutions to what were once live research programs. The piling up of settled problems does not, as noted earlier, provide a general definition of progress, since the percentage and significance of solved problems in a field require assessment, and there is the problem, waived as a presupposition here, that what is settled may become unsettled when new data, new instruments, or new theory attacks the convergence of language and data in some area of science. Kuhn's account misses the impact of new instrumentation and new data, which can create a new data domain whose adapted theoretical form will typically be obtained by sudden crossbreeding from the existent stock of theoretical ideas.

Popper is wrong to see permanent revolution as the goal of science.

The instrumentality of science to solve new problems may harden if new domains and theories for them are not developed, but settled domains are a necessary and desirable feature of scientific history. If new human relationships require new literary invention, it does not follow that definitive treatments of such staples of human interaction as anger, lust, and sorrow do not exist in older literature that we should want to keep. Similarly in science, where various practical problems remain common to our past, and for which appropriate scientific knowledge exists.

What individuates research programs and their theories is contained in the notion of data domains. For this reason, it is not necessary to postulate, as Lakatos does, a fixed hard core defining theoretical identity. All of the researchers attempting to fit a given domain may evaluate the conscious theoretical formulations differently, one researcher seeing as central what another holds open to debate. This possibility seems more closely instanced in the actual divergent opinions of scientists than the notion of a paradigm or hard core suggests, but it requires a relational view of theory and domain. Further, research programs may progress or degenerate quite independently of their final significance in scientific inquiry. A research program solving fewer and fewer problems may be approaching appropriate consonance of language and data in some important area, or it may be unable to adapt to relevant data produced by a new or refined instrument. Finally, a call for new theories and new research programs is somewhat quixotic in the absence of domains to be filled. A more desirable call is for preserving diversity of existent theoretical variants over known data for the purpose of maximizing the recombinant possibilities of theoretical concepts known to be viable when data demands new theory.

The situation to be discussed in science can be seen most clearly in mathematics. We can regard the various number systems of mathematics as data domains to which theories have been adapted. For simplicity, such further domains as vector spaces, matrices, geometric planes, rings, and fields will be omitted from this discussion, but similar observations would be relevant. For example, for the positive integers the number of prime numbers, the number of twin primes, the truth of the Goldbach conjecture, and so forth have been matters of theoretical dispute, some of them settled. In contrast to science, the data domains of mathematics are perfectly precise, and they are not subject to significant perturbation after they have been clearly marked out. Further, many theoretical claims about such domains are either true or false, and no real dubiety about the matter exists after the construction of appropriate proof. As a result, the logic of falsification works quite neatly, and significance is frequently obvious (because related only to truth or falsity in a domain) at the time when various proposals are put forward. Positivists and others have frequently treated science as though it were like mathematics, with precise domains and facts within them, and with clear methods for determining truth or falsity. Positivism might have provided a satisfactory epistemology for science if science were like mathematics in terms of having clear data, the significance of which is resistant to change over time.

The instrument of mathematics is proof, and this instrument has been in use, although with refinements, at least since Pythagoras, Euclid, and the other Greek mathematicians in the relevant tradition leading to modern science. As in the scientific case, instrument and data domain can interact, as is shown brilliantly in Lakatos's study of the interaction between proof and counterexample leading to a settled concept of polyhedron.[21] In Euclid's work, the extent of plane geometry and the domain of positive integers are not quite coincident with modern notions related to axiomatics and formal proof, but these data domains have shown a remarkable stability, and for the most part Euclidean proof is easily adapted to modern standards of rigor. There are no continuous data domains of this longevity in the scientific tradition. It is not surprising that mathematics and astronomy should have developed early in our view, since their pursuit, at least initially, was not dependent on sophisticated instrumentation.[22] The excitement of mathematics for a philosopher like Plato must have been the first adumbration of the organic theoretical growth of scientific theory based on settled foundation in fact. Plato's skepticism about empirical theoretical knowledge was appropriate until the rise of an instrumentation that could permit legitimated fact to arise.

In mathematics, data domains are constructed or invented by the human mind. Provided that methods of proof for extending these domains do not encounter inconsistency, there is nothing that could change them. Invented domains can have very complicated structures, with various notions of proof exposing different structure within the domains. Domains may wax and wane in terms of their interest to mathematicians, and domains may be shown to have the same structure or some other logical relationship that is of significance for methods of proof and the typology of mathematics, but certain kinds of surprises can be ruled out. Imagine the discovery that a mistake has been made all along, and that there is no fourth positive integer. By contrast, the significance of fact in the data domains of empirical science is subject to revolutionary shift with the discovery of new data. Entities thought to be identical may diverge, and vice versa, and the discovery of new entities may completely change the reference class of possibilities for what was already thought to be understood. Of course, even mathematical ideas and domains like the positive integers need not be given full blown. As Klein and Bloor have noted, Greek conceptions of number, and our own, are somewhat divergent.[23] To obtain the structure of the positive integers as a fixed data domain, decisions must be made about the first integer (is it a divisor, and is it one of the *positive* numbers?) and about the infinitude of the collection, and these decisions will interact with simplicity of proof and theorem, the acceptance of mathematical induction as a valid form of proof, and so on. Yet the structure of the positive integers can be discerned in Euclid in much the modern form, and although extensions of this structure to the full number line occupied mathematicians for twenty centuries and more, theorems about the structure of the positive integers have remained intact throughout this period.[24] There is currently little research on the positive integers. Although not all problems about this structure have been solved (the Goldbach conjecture, Fermat's last theorem), it exists in solid analogy to the settled domains of science discussed earlier. The permanence of data domains, once appropriately constituted in harmony with accepted standards and styles of proof, seems the major touchstone for understanding the differences between mathematics and science that have intrigued a number of scholars. This permanence, related as it is to human invention, helps to underscore the importance of instrumentation in enabling the continuous creation of new data text in the empirical sciences. Mathematics must deal with the data domains found intelligible and consistent by human beings. Their relative simplicity and frequent recurrent features, and the difficulty in inventing genuinely new and useful mathematical data domains, underscore the relative lack of complexity of inventions of the human mind in contrast to the constant frustration imposed on human theorizing over reality, the unanticipated richness of whose structure has subverted every known attempt at consistent conceptual description in other than limited and partially abstracted domains.

Some theorists have drawn a sharp distinction between the arts (and also possibly law, theology, and the humanities) and science, and have then attempted to place mathematics in between, as sharing some features of both science and art.[25] When science and art show too many similarities, it is assumed that something must be wrong with the given methods of analysis, since the distinction between them is taken as an important phenomenological given. But it is simply possible to consider painting, for example, as a series of styles governed by shared outlooks and attacking various technical problems in much the manner of scientists in history.[26] In this view, shifts in style may look like shifts in paradigms or research programs, and painting may exhibit a diffuse notion of progress not unlike that which we have been grappling with in science. In all of these areas, the successes of the past can be incorporated into the vocabulary and techniques of the present, so that it is possible to be a contemporary practitioner without detailed historical knowledge of one's discipline. Nonetheless, knowledge of the history of one's discipline may be most useful to artists and humanists, less useful to mathematicians, and least useful to scientists. Are we then dealing with some sort of a continuum?

What seems to matter most is not the relevance of history, but the existence of data domains and theorizing about these domains. It is being suggested here that mathematics and science form a continuum in this dimension in which the range is specified by precision and permanence of data, coupled with the possibility of permanent extension of data text. Between science and mathematics will be found abstract theorizing over ideal types, objects not found in the real world, such as perfect gases and frictionless surfaces, but objects sharing the dimensions of real world objects, with simplified properties and subject to idealized instrumental investigation yielding precise data similar to that obtained in mathematics. Mathematics shares with science, and shares most closely with idealized science, accuracy of theoretical prediction and the possibility of refutation of prediction through the accumulation of data. By contrast, the arts and humanities deal with the description and interpretation of relatively fixed texts, and with problems capable of expression and resolution within the limits of ordinary language and experience. Ordinary language and experience may be stretched or extended in the arts, but it is not augmented into a language for specialists. This observation, of course, would require modification for recent extensions of the humanities that have been influenced by the style of the sciences, and have searched for autonomous subject matter and autonomous languages and relevant evaluation only through expertise, but the origin of the sharp phenomenological distinction between science and art must lie in attitudes formed before these recent and quite problematical developments. We can consider a few examples. Western painting has represented worlds of general human interest (albeit in a succession of styles) through similar technical means for centuries, and has only recently moved in many quarters away from traditional anthropomorphic subject matter.[27] In this tradition, the human figure may be described and rendered from many points of view, all of which are consistent within the confines of relative description. There is no analogue to prediction and refutation, save where a theorist or painter contrives to predict the direction of painting or the public response to some style. In theology, the texts to be interpreted are notoriously set in advance through the appearance of God or prophet in human history. Although interpretation must be made contemporary to be credible, such general problems as the distribution of evil in God's kingdom can remain problems of interpretation of the same text for centuries, and the notion of the instrumental augmentation of data achieves no purchase in this situation. Whatever the complexities involved in relating science and art, it seems likely that mathematics will always show close differential affinities to science, and will remain bound to science in such discussions through the existence of mathematical scientific theories interpreting idealized data that characterize postulated idealized objects. The differences between mathematics and science are to be related to the consonance of theory and data more easily achieved in mathematics, and the permanence of that consonance as a result of the invented nature of the data domains. What has been noted as a difference in history is then primarily to be explained as follows: Some domains and theories achieve a perfect fit in mathematics, and then no longer change over time, no matter how useful the theories may be. Research interests then shift. Mathematical systems, which develop an invented domain, can be terminated by inconsistency, or have their range restricted by impossibility proofs, but they are not so subject to counterexample by unexpected data. Unexpected data cause revision, but also lead to retention of what has been shown, quite in accord with the classical positivist view of scientific history. By contrast, domains in science are likely to be deeply reevaluated when later instrumental augmentation of data text produces totally unanticipated consequences. Fewer settled theories of any longevity are to be expected in scientific history.

Kuhn, in a recent discussion, compares mathematics to art: While stagnating traditions die out in each, stagnating traditions may be preserved for their utility in application in science.[28] In this connection he cites two interesting papers by Fisher on the death of a mathematical theory, specifically invariant theory.[29] Although Fisher discusses what he sees as differences between some of the sciences and mathematics in his articles, this pairing by Kuhn is somewhat misleading. Perhaps no one paints seriously in certain past styles in the present. The number of invariant theorists is not quite so small.[30] Our textual comparison between science and mathematics results in the expectation that the distribution of mathematicians would change more dramatically than the distribution of scientists after consonance between language and data, or after the appearance of counterexample. In Fisher's view, the problems of invariant theory were to a large extent solved by a famous and unexpected proof by Hilbert, greatly reducing its scope, and it was reduced to nearly total stagnation as interests shifted to vector and tensor theory, as well as to axiomatics as a proof style, in the early decades of this century. The reason for sharp shifts in mathematics is the relative clarity of the logic of mathematical proof and counterexample as a result of the precision of both theoretical consequence and data. Mathematics may mimic art in stagnation, but the mechanisms are different. Art interests shift with saturation of style and boredom with a technique, whereas mathematical interests shift when challenging problems are solved. Further, contrary to both Kuhn's and Fisher's impressions on this point, it seems that essentially solved domains are retained in both mathematics and science, although they will be more numerous in mathematics.[31]

After consonance of language and fact is achieved in science or mathematics through social legitimation, it may seem quite marvelous that language should fit reality so closely. The usual correspondence theories of truth trade on the established fit between ordinary language and the world, or between contrived calculus and idealized domain. In the case of ordinary language, a close fit in many areas was achieved long in the past and the amount of noticeable current change is small. Because of this achieved fit, it is rare that one cannot describe and evaluate, for ordinary purposes, whatever one confronts, at least as far as the limits of language are concerned. New words frequently seem to be only abbreviations for features of the world that can already be described with greater circumlocution in the linguistic resources already available. In mathematics and science, the process of achieving this consonance is more open to inspection, since new data domains and linguistic resources to describe the domains are always undergoing a forced and painful development. Once clear statements of fact have been achieved through instrumental investigation, the reference of fact seems fixed and objective, and indeed it is. The world has been discovered to show a fixed and repeatable response in certain interactions as described in the language, and this response is an objective consequence of these interactions. Even mathematics and scientific theorizing, like certain other forms of fiction, can be viewed as limiting cases of this generality, where the response to interaction is a highly idealized version of actual responses designed to preserve features of simplicity in the language used to describe that response. This process of achieving or constructing reference for language by development of a domain we will call the microprocessing of fact, after discussion of this phenomenon by Latour and Woolgar.[32] When the process is complete, the evidence of microprocessing disappears, and mere correspondence, the very correspondence that has been slowly and carefully constructed, is all that remains.

We can start with a mathematical case, by considering the concepts of integer and set. The domain of integers has been investigated for more than twenty centuries, and was adumbrated earlier in the process of counting. What an integer is seems clear and straightforward, and whether or not something is an integer is open to a principled inspection in the relevant mathematical cases. There is an identifiable zero, and all the other integers can be reached from this zero by a simple operation. The relationships of positive integers, negative integers, odd integers, even integers, prime numbers, and so on are so well understood that the domain has a clear outline, clear methods of proof, and only a few unsolved problems. Terms such as *positive integer* have clear reference, and code into their significance the resolution of disputes that were live in Greek times, but have been settled for centuries. For example, the term *integer*, in contrast to the term *number*, literally codes the generally recognized fact that the integers do not exhaust the numbers, and so on. There *are* integers. This much is beyond dispute even though there are various philosophical explications of the implied ontology. When we turn to sets, everything changes. The total domain of sets has been explored for less than a century, and disputes into its extent and nature remain unresolved. Various axiomatic treatments of set theory, for example, differ in the language that they use, and even differ over whether certain sets belong to the appropriate domain of investigation. In comparison to the concept of an integer, the concept of a set seems fuzzy, except to some partisans of fixed approaches. The social process of negotiation here is still so obvious that some expert mathematicians believe that the paradoxes and problems of set theory will never be totally resolved, so that set theory remains a useful tactical instrument, but is not the appropriate theory for discussing foundations of mathematics along the lines that have been pursued by set theorists in the first half of the twentieth century. The microprocessing of set theoretic facts has not been completed, and the reference of set theoretic language remains somewhat shifting and uncertain beyond various simple consensual examples. Whether there is a clear domain of all sets and an appropriately sharp ontology referencing set theoretic language remains an open question.

Between mathematics and the experimental investigation of reality lies an area of mathematical theorizing in which empirical content can be simplified and greatly extended by taking on a mathematical form. Here we find idealized objects capable of mathematical description and mathematical development in theory. Typically, the empirical data will resist idealization, but a fit between mathematics and reality of sufficient closeness can sometimes be forcibly achieved that allows mathematical anticipation of reality, allows a way of anticipating pattern in data.[33] The planets do not behave exactly as in Newton's calculations, nor do gases as in Boyle's, but the mathematical theories of both anticipate fact that could not be gathered from mere accumulation of data. We have noted the descriptive fit between ordinary language and reality, but there is a predictive gap. The weather can be described, but not predicted reliably according to a theory expressed in ordinary language, as well as the fact that comets can be described, but not the timing of their appearances, and so forth. Predictive gaps can be filled in by constructing a mathematical model of reality and then predicting the nature of idealized events within the model. Slippage between precise prediction and observed fact can then be used to fine tune the model, and so forth. There is no satisfactory philosophical explanation of the widespread success of this process in science, but its dialectical contribution to scientific advance is part of the epistemological description of scientific knowledge being urged in our study.[34] Some of the mystery is removed by seeing this dialectic as existing between pure mathematics and experimental inquiry. Pure mathematics deals with even more idealized data. Objects are, for example, to be merely counted, and mathematics can allow us to predict the result of a total count by summing partial counts. Where this is successful, and we must learn through experience where it is successful, we can utilize mathematics to avoid the wholesale translocation of objects in the world to obtain a full count, even though objects are hardly as neatly individualized in the real world as they are in mathematical representation. Exactly how the mind can match mathematics and empirical fact remains a puzzle, but the matching is the secret of scientific epistemology and its dialectical progress. Discoveries such as that of Neptune by Adams and Leverrier depend on the supposition that theoretical and empirical description can be brought into arbitrarily close coincidence through refined instrumentation coupled with sufficiently clever mathematical idealization.[35] Had Adams and Leverrier maintained a Platonic separation of world of being from world of becoming, looking at a specific place for a specific observation could not have confirmed or refuted any sound theoretical position. The role of instrumentation in bringing coincidence of theoretical and observational language, and the ability of modern mathematicians to construct idealized mathematical representations of reality for test through instrumental observation, are jointly responsible for the explosion of scientific knowledge since the seventeenth century.

The dialectical interplay of fact and theory in this vast middle ground of scientific theorizing depends on the existence of sufficient legitimated fact to give a domain stability, and to provide a language in which the domain can be described. As in the mathematical case, legitimation of fact involves modalities and obscure reference. Scientists are interested to know whether their statements refer to facts "out there," or reference mere artifacts of experimental or theoretical technique.[36] During legitimation, investigators may be idealists, relativists, skeptics, realists, and so forth, as the evidence shifts with respect to their scientific intuitions. Genuine epistemological debate will fuel the controversies required to develop the relevant data domains. The factual status of a substance will shift along with the modifiers required in its description by evidence. A better language, as well as better instrumentation, will be a desideratum of investigation. When consensus is reached, it is not surprising that language and fact will mesh. Language will contain terms for the facts that have stabilized in the investigation, and the facts that have stabilized will call for a referencing term in the language used to describe the domain of which they are a part. Statement now matches reality because they are two aspects of the result of the same process of exploration and negotiation. It is only after statements match reality that they can be "about" it in the dialectic of science. And the fact and the language may melt once again into indeterminacy as the terrain of domains involved in science shifts. A term like *gene* was once obscure, then clear, and then obscure again, as biology developed. A term like *ether* was once clear, then obscure, and then clear again, as cosmology developed. As we move to empirical science, the possibility of cycles of significance arises in a manner not anticipated in everyday life or mathematics, where terms usually only acquire significance and then retain it, although terms may become obsolete as interests and style of life undergo a change.

What we have called microprocessing is accelerated by the scientific method. Instrument, language, and fact can relatively quickly achieve consonance, and retain that consonance unless new instruments or new data upset the harmony of their relationship. We have already mentioned some relevant examples. Millikan's experiment quickly established the existence of the unit charge and the indivisible electron, and even if the exact value of that charge has shifted a little with more sophisticated techniques, the existence and approximate value of the unit charge have been established facts for nearly a century. In Latour and Woolgar's discussion of the discovery of a releasing factor, the existence and structure of the factor were quickly settled when a method of obtaining appropriate spectrometer data was found. We have noted that Barkla's J phenomena were not accepted because other physicists looked askance at his experimental methods involving heterogeneous rays, and preferred to develop a data domain revealed by a spectrometer utilizing homogeneous rays. A study of chemical theory culminating in modern valence theory by Gay indicates that the relevant data domain was discovered and stabilized by the experimental use of the voltaic pile and platinum crucibles, leading to the resolution of divergent theoretical attitudes and the creation of modern chemical ideas about the bonding of molecules.[37] Fleck's study of syphilis and the Wassermann reaction is perhaps the most revealing, largely because of the vagueness of the concept of a disease that can at first be known only through the presence of highly variable symptoms.[38] The concept of syphilis as a disease entity, as well as its relationship to other disease entities, was considerably complicated in its history by the disease's relationship to moral attitudes about the mode of its acquisition. It is also interesting to note that the discovery of the relevant microbe did not settle the nature and existence of the disease, since the presence of the microbe is related to symptomatology in an extremely complicated way. The Wassermann test and the emergence of serology were decisive in the context of medical investigation looking for the existence of blood factors. This ultimately permitted an intelligible relationship between acquisition, treatment, and cure to exist as a medical fact. Wassermann, of course, had not set out to find the test that has given him a place in scientific history, although he later interpreted his work as directed at that goal. The goal could not have been coherently described when Wassermann began. Fleck compares this to the voyage of Columbus, who sailed for India but discovered America. Language and reality then required (in both cases) severe adjustment. In these examples, we are confronted with instances of the consonance of language and fact that have remained stable since their occurrence.

It seems reasonable to conclude this section with a rather extended example of fit between instrument, domain, fact, and language that has remained stable in the face of subsequent revolution in theory. The longevity of Newton's achievements, many of which have remained stable in the face of considerable elaboration, will suffice. We will take as our central example the area of physics known as classical mechanics. As originally developed by Newton, this is perhaps the first mathematical theory of nature to promise organization and explanation of a potentially infinite number of experimental findings.[39] Until the late nineteenth century, Newtonian mechanics was equivalent to exact empirical science. Since then, Maxwell's electromagnetic theory, relativity theory, and quantum theory (among others) have rocked physical speculation, but they have not overthrown classical mechanics, although they have forced a much more limited view of its domain. It is still taught to students of physics, and deserves to be taught, because it is a correct description of nature in those domains where its concepts fit facts. The newer theories belong to domains where instrumentation, fact, and theory are not consonant with classical mechanics. They have not driven classical mechanics out of its own domain; they show rather that classical mechanics is not viable in their domains. Heisenberg has called theories like classical mechanics closed theories.[40] They are consistent, and they are firmly anchored in data domains through appropriate instrumentation. As long as data to be described, explained, and predicted belong to the appropriate domain, the theory cannot be improved on, and it is best regarded as true. The domain reveals an aspect of reality, and displays that aspect in a set of related scientific facts. Should not the statements fitting these facts be regarded as true of these facts? The fact that they are incompatible with the facts in other domains is independent of this judgment, as is the observation that they may not fit their facts *exactly*. With this observation the clash between the use of classical mechanics within physics and the philosophical observation that more recent theories have shown classical mechanics to be false is resolved without normative consequences for scientific practice. A similar point could be made about other domains and the languages fitting their legitimated facts. Some scientists believe that quantum theory itself is closed, and that it has adapted so closely to the relevant facts that it must remain intact within future physical theories just as classical mechanics remains intact within physical theory.[41] The existence of closed theories is an important aspect of scientific progress, allowing adaptation to occur into new domains without jeopardizing the viability of theory within already explored domains. We can see it as dynamic equilibrium achieved in the dialectic of theory and experiment between theory and legitimated fact, an equilibrium achieved where domains have remained stable long enough to permit successful theoretical adaptation.

Theory and Experiment

In the last section we have discussed the situation that exists when a data domain stabilizes and consonance is reached between instrumentation, fact, language, and theory. It remains to discuss the situation that obtains when new theory or new data erupt so as to destroy consonance, or to provide the stimulus for an attempt at new consonance. Either theory or experiment can advance alone. When appropriate consonance is reached, neither the addition of data text nor refinement in theory need have a catalytic influence on their relationship. Here we get science from science within a relatively closed domain, theory elaborated from existing theory, or experiment from existing experiment, without the necessity for constant dialectical accommodation. A novel instrument or observation may suggest a new domain, calling for novel theory, or a novel theoretical idea may project a possibly novel range of data, calling for experiment and instrument for confirmation. At such growth points, there may be a limited but enormously stimulating contact between theory and experiment, in which one or the other tends to define a direction of development for the other. It is at such points, where science is not developing out of known science, that external factors such as philosophical or social influences may play their most dominant role. Fixed ideas are required for advance. Fleck has compared this to the motion and inhibition required to move a limb.[42] Perhaps this is why the biggest moments of change may be associated with the interaction of theory and experiment, in which a fixed projection from one requires drastic accommodation from the other. During consonance, the two factors can develop more or less independently toward maximally useful adaptation.

We refer here to a dialectical accommodation of theory and experiment that is most marked during disequilibrium. Although science as a whole may show traces of this process, it cannot be expected to reach into the careers of individual scientists. Some biographies will reveal marked turns as the result of the influence of an experiment or a theoretical concept, but many will show an unyielding trajectory. The process described here will then be evident in consensus judgment over time, but all sorts of individual inflexibility will also be evident. This complicated relationship between accommodation in consensus and deviancy in private biography is what by now we should expect. Pushing some experimental line or theoretical bent to its extreme may be a necessary part of determining the significance of the ideas involved, and the evolutionary model suggests that the appearance of such inflexibility in the distribution of attitudes plays an important role in determining the direction of group development. We can expect this feature to be most marked in physics, where the separation of the role of theorist from that of experimentalist is most clearly to be discerned.

Up to this point in our discussion, we have treated data domains as though they were isolated, and as though theories adapted to them in a one-to-one relationship. It is time to examine competition between theories for domains. Two or more theories may continue in a struggle for adaptation to a domain in which consonance is not reached. If the data do not change suddenly in nature, it is to be assumed that increasing the data text and experimenting with theory and interpretation of the data will eventually produce consonance. But it is the case that theories dependent on different concepts, such as wave and particle interpretations of light transmission and electron transmission, can exist side by side for extended periods of time, each style of theory explaining some of the legitimated fact and predicting similar fact, but unable to deal with the entire data domain, or unable to deal with it with the accuracy displayed by the other approach with respect to some of the data. The competing theories are likely to prefer different experimental arrangements as fundamental, and to regard different selections of fact as most important in setting a theoretical treatment of the total domain. Under these circumstances, adaptation is similar to that in certain biological niches, where closely related organisms eating slightly different food and distinguished in terms of behavioral repertoire share a niche, each concentrating in numbers at its maximal fit, and decreasing as one moves away from this optimal point. Its share in the total distribution may become small, but a viable variant remains ready to rush back into larger portions of the data domain should the extent or significance of the facts in the domain suddenly alter. Variants, so to speak, become consonant with parts of the domains, and are as well adapted to these parts as any theory could be. Provided that at least some significant problems depend on data only from such subdomains, the theory will remain viable, even if in this version it is not important in majority scientific discussion.

At times in the history of science, a general theory will systematize, reorganize, and provide significance for what has been a range of more specific adaptations to local data domains. The great theories—classical mechanics, electromagnetic theory, relativity theory, quantum theory, statistical mechanics, valence theory, Neo-Darwinian evolutionary theory—have all accomplished this. In many cases, the general theory will not make exactly the same predictions from given data, or explain the data in the same way as the more restricted theory, but the general one has the advantage of being viable in a much larger domain than the more restricted theory. Here we encounter a point where the analogy between theory and species does not hold. Biological species consume food in order to maintain life as long as possible and to reproduce their kind. Sometimes, however, food for a species may consist of other species, as in the predator and prey relationship. Theories explain data and predict data. They do not consume other theories, and no analogue to the predator and prey relationship seems to exist. The facts in data domains are not strictly analogous to bits of food, nor data domains to ecological niches, nor theories to species, even though this evolutionary metaphor is of considerable use in considering the nature of scientific knowledge and its growth over time. A theory is a disembodied set of response mechanisms, analogous to an organism that does not eat, but merely adapts to stimuli. It is most closely analogous to the neural adaptive capacities of organisms in their response to environmental input. This is the source of value in the analogy that we have been pursuing.

Theories are strictly analogous to adaptive mechanisms that human beings can exploit in fitting and extending their ecological niche. As such, they may be compared to patterns of behavior that human beings can learn, thus increasing the sophistication of human response to stimuli. When human beings learn more sophisticated responses, their neural pathways will be altered and augmented. Theories can't be ontologically grounded as neural pathways, since theories, along with most cultural artifacts, may exist in books, films, photographs, and other forms that are not internalized but can be consulted when required by humans. Just enough may be internalized to cause the human being to remember where to find help and to utilize that help. We can regard theories as extensions of human thought, just as instruments are extensions of the human sensory apparatus. As with other thoughts, the assertions in theories will be true or false relative to their domains. Different human beings beings will have different perspectives on externalized theory, and may internalize slightly different modes of interacting with theory. These variations, which we have come to expect, may all prove adaptive in the given data domain, while preserving alternative possibilities for adaptation in new domains. The final ontology of theory is as difficult as the ontology of species. It must comprise a variety of response possibilities, not all of which need be in existence, and these possibilities will share both a style and a range of stimuli. Discussion of a theory, as opposed to some individual variant of a theory, is similar to discussions in ethology of the typical behavior patterns of a species, which may not have a precise instance in any particular organism. The ontology of theories is as complex as that of many abstract objects, but we can make do by distinguishing theories in terms of their languages and assertions, which explain the response style, and the data to which they are directed, which explain when their response patterns will be engaged.

For simplicity of exposition, we can imagine that a scientist has internalized all the scientific theory he or she knows, so that it is all represented in his or her neural circuitry. The usefulness of storing memory and technique externally, or even its practical necessity, will then not needlessly complicate the relationship of theories. As human response becomes more sophisticated, reflex may be replaced by crude learned behavior, and the latter by increasingly refined learned behavior. One response pattern doesn't eat another; it can be used instead. This is why predacity is not a feature of theory formation. Reflexes and cruder patterns of response may need to be replaced by more sophisticated learned behavior when anticipated results do not obtain, or when data become more sophisticated. New patterns of response obtained by practicing older patterns may call out for appropriate stimuli. In this process of learning, the older response patterns will remain in existence. The older response patterns do match reality, and in at least some cases reflex may be preferable to cognition. Cognition serves to bring together data from a variety of sensory modalities and test their significance against stored information and existent response mechanisms. Older response patterns will remain in some form in the pathways of most sophisticated cognition, because the old pathways will have proven value in various situations. New theories contain incorporated but modified parts of older theory. The advantage of general theory is in sophistication of response to a wider possible range of data, responses being modifed by information from this wider range. Where the data are gathered by the most complex instrumentation, the wider theory will be preferable in use, and the older pathways will be circumvented.

Reduction in the philosophical sense is a way of finding formal relationships between theories that have received an axiomatic form. The simple historical fact is that reduction has occurred in the philosophical sense very rarely, if at all, in the history of science. An older theory gets incorporated, with modifications, into a newer theory, or attenuates and dies out as a response probability. For various data, two response possibilities may very nearly coincide in their anticipations, but they will be different, as a reflex is from learned behavior. If data domains can be explored, extended, and carved out anew, as we have argued, then reduction to one universal theory is a practical impossibility. The goal of unified science seems to be a legacy of theology, and the view that there is one way the world is.[43] In theology, God is sometimes said to have created the world according to a blueprint, which, if it could be known, would explain the precise structure and workings of the world. The idea that physics, in pursuit of the smallest constituents of reality, is uncovering the fine-grained blueprint, while the methods of other sciences provide convenient shortcuts, appears to be the descendant of this outlook. In the view offered here, the complexity of reality defeats the unified science hypothesis, although the drive for more general theories remains a desirable aspect of scientific motivation. Different sciences will offer different modalities of interaction, all of them equally legitimate. This does not cut against, or for, religion, since God can be imagined to have created a world not fully within His rational grasp, although this requires some adjustment of the concepts of His omnipotence and omniscience.[44] The most frequently cited reduction in scientific history, that of phenomenological thermodynamics to statistical mechanics, illustrates our claims. As is well known, the temporal asymmetry of phenomenology disappeared in this reduction, even though many of the other statements in phenomenological thermodynamics were taken over wholesale. This is the pattern we expect. A new theory will incorporate major features of older theories, but it will modify some features in extending its applicability to a wider domain of data.

Let us consider a new data domain, perhaps one being explored with new instrumentation. At first, older theoretical responses will be tried. If the older responses are not adequate because they do not fit the structure of the data, or are predictive of nonobserved data, they will typically be modified by mixture of older response types. This is similar to biological adaptation from existing stock. Some lucky mutation can occur that is advantageous, but the standard method of adaptation is to shuffle existent genetic possibilities in new combinations. (This shuffling, of course, constantly goes on, but it doesn't shift distribution of genotypes dramatically in a *fixed* environment.) New data are always both like and unlike existent data in an infinite number of possible ways. The yardstick for its evaluation must be existent language if chaos is not to reign. Goodman proposed that entrenched classes of objects be used to project new data as a way of avoiding predictive chaos.[45] This proposal deals with the problem of induction by arguing, not that the relationship of past to future can be guaranteed in methodology by accepting that theory most likely to be true, but that we always try existing projections until they encounter difficulty. Holton has suggested that scientists theorize in terms of themata, that is, in terms of a set of ideas that are highly adaptable to data, such as viewing various entities either as waves or as particles, and so on. These concepts, which have a long history in our thought patterns, can be reshuffled and refined to produce a new theory when data so demand.[46] Holton's view allows new theory to arise from crossbreeding between older theoretical conceptions, and permits an extension of Goodman's idea to situations with complex conceptual antecedents and to situations in which divergence between theoretical outlooks is possible. According to both of these accounts, new data do not create new theory; they create primarily reshuffling of existent theory to find a new language. When this process leads to a quite new language in a short period of time, a revolution may seem to have occurred. When a domain is rapidly expanding, divergent theories or research programs may be simultaneously progressing, just as rival colonizing species may all be increasing in number in a new ecological niche. The mere fact of progress in solving problems and accommodating data may not be an indicator of eventual success. As we have noted, rivalry may exist for a long period of time, or be terminated by extension of data text or the production of a more general theory. Many commentators have noted the widespread appearance of analogy and metaphor in scientific understanding. From the prospective suggested here, analogy and metaphor are the mechanisms by which the style of older theorizing is extended to new domains, and the two would be encountered wherever data are causing theoretical readjustment.[47] The much discussed simplicity of nature may be the reflection of the fact that reshuffling of theory has proved such a successful strategy to date. New data domains have always yielded to a similar means of theoretical extension.

In the biological domain, reshuffling of genetic material is primarily the result of a random or pseudorandom process in meiosis. If methodology could be made explicit, or there were an inductive logic, no room for reshuffling would be available at crucial junctures. In science, the fact that individual scientists vary in personality, aesthetic vision, and so forth provides room for a variety of possible reshufflings of existent theory (and allows key mutation), the better variants of which process will be determined by group interaction with the relevant data domain. One method of extension can be to reevaluate existing theory and to find a general theory that fits the important cases in existing theory, although it condemns other data to the status of being wrong or irrelevant. Newton, in generalizing from Kepler's laws, took them to be important in a way that neither Galileo nor Descartes could have, in view of their settled interests in the Copernican and vortex theories.[48] In adding the notion of mass to that of inverse attraction, Newton was able to incorporate slightly altered versions of Kepler's laws in his theory, while decisively repudiating versions of facts accepted by Galileo and Descartes. Another method, particularly in the case of complex data, is to allow an image from common sense to organize the data until a mathematical refinement makes scientific sense. Gruber has shown how the image of a branching tree allowed Darwin to suspect a pattern behind the complex data of breeder's observations, and so on, until a suitably precise scientific form of that image could be constructed from existent concepts.[49] In this image, there is continuity in branching, but the termination of a branch marks cessation of growth in that area. Extinction of possibilities is thus required if the total number of existent organisms is not to grow beyond observed limits. Much of the imagery to be made precise in the theory of evolution is contained in this general image. Wise has shown that an image of similar generality, consisting of electric and magnetic current as represented in a pair of interlocking rings derived from Faraday, guided Maxwell's development of the mathematical description of the interrelation of electricity and magnetism.[50] In order to consider the reshuffling of themata along the lines suggested above, we can turn briefly to the history of quantum theory from 1913 to 1927.

The development of quantum theory from 1913 to 1927 not only shows reshuffling of themata, but also shows how different scientists attacking the same data may piece together different theoretical advances by means of choices among existing possibilities.[51] In this process, the visualization of nature's structure that classical theory allowed was destroyed, and then regained as the fit of quantum theory to the data allowed scientists once again to feel that they could see structure by means of their theory. In 1913, Bohr proposed a new theory for the classical atom that retained Rutherford's picture of the atom as a miniature solar system, but involved open conflict with classical physics in its consequences. Through a series of small steps, Bohr was forced to yield Rutherford's picture while adopting a nearly incomprehensible theory welding both continuous radiation fields and discrete particles of matter into an explanatory apparatus for the accumulating data. In the ensuing development of theory, choices had to be made between the themata of continuity and discontinuity, wave and particle, causation and noncausal "jumps," and between mathematical models of process with an uncertain relationship to reality and more familiar mechanical models. By mid-1925, Bohr and Heisenberg were forced by their interpretation of data and choices in favor of discontinuity, particles, noncausality, and mathematical models to develop an essentially unvisualizable mathematical model of quantum processes to fit the important experimental data. In 1926, Schrödinger made choices in favor of continuity, waves, causality, and realizable models, and developed a wave interpretation of the same phenomena while stressing slightly different experimental data. Shortly thereafter, a somewhat eclectic interpretation was developed by Born in which the particles were seen as guided by waves, and partial visualization was restored. Controversy between the various theoretical outlooks, with unmistakably fruitful consequences for quantum theory, was partially resolved by the discovery that wave and quantum mechanics were essentially intertranslatable pictures, allowing one or the other to be used in conformity with the needs of any special problem and the expertise of the individual scientist. In 1927, Bohr presented a theory that resulted in the so-called Copenhagen interpretation of quantum theory, which a majority of physicists soon accepted. The particular example of quantum theory shows that alternative shufflings can lead to equally adapted forms of theory. In other cases, of course, nonequivalence of reshuffled theory will lead to a markedly skewed distribution of individual interpretations in the light of evidence.

To this point, we have viewed new data as the primary motor of scientific development, and we have examined possible lines of theoretical adaptation to new data. In order to sustain a dialectical interpretation of progress, it is also necessary to consider the way in which new theory can force a recognition that new data are requisite. The most obvious mechanism for this is well known. Development of an existing theory can lead to the anticipation of data to test its fitness. The discovery of Neptune and the discovery of the positron are examples of this mechanism, involving somewhat different levels of data augmentation.[52] Existing theories can also be discovered to conflict with future data when they are developed, thus leading to the possibility of an experiment that can decide between them. A more interesting development is when theoretical reshuffling produces such a conflict, a situation that can prove stimulating to experimental development. An interesting example of this phenomenon is the controversy that raged in cosmology for some time between steady-state and evolutionary models until the data produced by the debate settled the controversy pretty much in favor of evolutionary models. In this case, the steady-state theory was produced, not by new data, but by a new interpretation of data in which theory was reshuffled because of consideration of the philosophical principle that the universe must be the same in its major features as one journeys out from our viewpoint in space and time.[53] Since local data would be equivalent on the two theories, it was necessary to invent new experimentation in the hope of adjudicating between them. In this case, of course, the theories did not turn out to be equivalent, and one of them had to give way.

The fact that some theories can replace other theories, or eliminate them from science, may seem incompatible with the idea that any theory that achieves community acceptance at some time in the development of science captures an aspect of reality. Phlogiston theory and steady-state theory did capture aspects of reality, but we regard them now as false because their explanatory claims diverge from fact in later, expanded data domains. As chemistry and astrophysics developed, the explanatory capacity of these theories lost out to superior alternatives. What they tell us of reality is therefore partial, and there is no reason not to call them false in the wider domains. At the same time, classical physics and static earth astronomy are still taught and still used, even though they have given way to other theories in expanded domains. Why are these different? Why are they still considered true, even if only in these restricted domains? Because the domains to which they adapt still retain an identity in science because of their close relationship to ordinary naked eye observation and purposes related to this fact that remain essential for human beings, such as the construction of machines and the navigation of boats. There must be both a separate data domain and an ongoing purpose for this domain if an older theory is to be retained and taught. If the wider domain to which relativity theory adapts had not been developed by new instrumentation, classical physics would still be adequate. The slight theoretical superiority of prediction of relativity theory within the domain of classical physics is compensated for by a more complicated and inaccessible mathematical apparatus. But since both are well adapted to the restricted domain as developed by the appropriate instrumentation, neither need force the other out. In the case of phlogiston theory, by contrast, new chemical data gathered by essentially the same sort of chemical experiment eroded the boundary and the legitimacy of the domain to which phlogiston theory had adapted. The ontology of rejected theories in larger domains is thus sensitive in its ongoing acceptance to the details of development of data domains and the continued usefulness of relatively natural domain boundaries. Our normal discussions of the truth or falsity of theories would need to be brought into coincidence with the historical details of domain enlargement and discovery of new domains, but the rather abstract treatment here gives promise that these accommodations can be achieved without violence to scientific intuition in specific cases.

Besides predicting new data, theory shift can cause a reinterpretation of older data, which in turn can stimulate the search for new data. One amusing example of this began with Thomas's nearly accidental discovery that injections of a substance called papain uniformly caused the ears of experimental rabbits to droop, a striking and regular change.[54] His examination of the cells of the ears of the injected rabbits showed no abnormalities, and he was unable to find an explanation. This experiment arose in the context of another line of inquiry. Some years later, when Thomas was teaching pathology, he showed his students the effect of the drug, and then examined the tissue of rabbits who were and were not injected. In this circumstance, the fact that gross quantitative changes in cartilage cells had occurred was obvious, even though none of the cartilage cells by itself showed abnormalities. This experimental discovery changed the prior theoretical opinion that cartilage was inert tissue and played no role in physical reactions. We see here how an observation in the right theoretical context can run counter to the theoretical background that would otherwise constrain its interpretation.

Another interesting example of theory shift causing a new interpretation comes from medical history.[55] At the beginning of the nineteenth century, scurvy and rickets had been discovered to be diet related, and dietary cures for the diseases were well known. In the historical context, the diseases were consequently thought to be understood. During the nineteenth century, the development of the germ theory of disease promised a universal explanation of disease as the result of infection by microorganisms. About 1880, beriberi began to be of interest to Western doctors. The fact that beriberi could be contracted by persons eating a wide variety of foods, that those who contracted beriberi might be on a poor protein or high protein diet, and other observations, convinced most Western researchers that the disease was not related to diet. Researchers looked for a microorganism as the causative agent. The breakthrough observation was that consumption of polished rice was responsible for a similar disease in chickens. In the new setting, the connection between polished rice and beriberi was not viewed as giving an understanding of beriberi as the connection between the absence of certain fruits and scurvy had done for scurvy earlier. If the polished rice did not cause beriberi through connection with a microorganism, it was necessary to find the factor (nutrient) whose absence caused the disease. In the new context, it became clear that the nutrients essential to health had not been purified and identified, and that this had to be done in order to understand deficiency diseases. Thus the existence of the germ theory gave the facts concerning diet and beriberi a new significance, and led ultimately to a modern deficiency theory of disease.

The result of this survey has been a welter of possible mechanisms for dialectical accommodation of theory and experimental fact. Experiment can lead theory, and vice versa. We should not be surprised to find that very complex interactions between new theory and new experiment are possible, ruling out any methodology based on either inductivist or deductivist proclivities. There can be no *a priori* mechanism for advance. One will have to be sensitive to the details of particular cases and the facts of any scientific setting. In order to clinch this observation, three detailed studies of scientific history of more than passing interest will be briefly described and commented on.

A study of the discovery of Boyle's law will constitute the first example.[56] Boyle's law was gradually formulated during a complex interaction of philosophical theorizing and experimental development in the seventeenth century. The interest of this case is that it stands on the borderline of an emerging scientific discipline where prescientific concepts play an important role and no settled experimental strategy is recognized. A flurry of European experimentation on the elasticity of air occurred from 1640 to 1650 as a result of a philosophical controversy between vacuists and plenists. Vacuists utilized the ancient intuitions of the atomists that motion was possible only if atoms could move within a void. Plenists owed their views to Aristotelian arguments against atomism, which asserted, among other things, that bodies passing through a void would all move at the same rate (indeed instantaneously), contrary to the observed difference in velocity between light and heavy bodies. When water was boiled and turned into steam, Aristotelians viewed this as an expansion of matter, which produced rarefaction, but not a void. Remnants of this dispute appear later in the quasi-philosophical arguments of Galileo, Descartes, and Newton. The consequence of this theoretical dispute was a series of attempts to produce a vacuum, the centering idea of which was to fill a tube closed at one end with water, to invert the tube in water, and to note that a space appeared within the tube over the water. If there was now nothing where there had been water, a void had been created. These early experiments were ambiguous because of the changing height of the water column, strange sounds observed during the formation of the supposed vacuum, and the embarrassing fact that the vacuum produced seemed to transmit light, magnetism, and sound. A series of counterexperiments was then located, attempting to prove that air (unlike water) was indefinitely elastic. In one such experiment, a carp bladder was inserted into a tube closed at one end and filled with water, and when the bladder was surrounded by the space over the water in the inverted tube, it expanded greatly (like a balloon), showing that a fixed quantity of air could fill a space of varying dimension. Needless to say, these experiments could be defended and attacked by both sides of the philosophical dispute in spite of efforts to obtain an unambiguous experiment to end the series. Here we have a very complicated interplay between rival theory and experiment.

Before long, the experiments referred to, largely conducted in France and Italy, had engaged the attention of British and German scientists. In England, Boyle constructed a pneumatic engine designed to pump air out of a container whose interior could be studied by various instruments. This sharp increase in instrumental sophistication was coupled with skepticism concerning the plenist and vacuist controversy. Boyle suspended judgment, thinking that neither side possessed a true physical description of air. A new series of experiments was then realized, for example, some experiments in which the density of air in the container, as measured by piston travel in the pneumatic engine, was plotted against the elasticity of air, as measured by the height of a mercury column it could support. Even prominent mathematicians could find no simple correlation, for the unknown fact was that the container leaked, and leaked worse as it was exhausted of air, and so the mercury column could not be accurately calibrated. Another series of experiments, performed by a variety of investigators but marred again by primitive equipment and uncertain calibration, did establish the diminution of atmospheric pressure with increasing height above sea level, a fact related to the elasticity of air. Not all of these experimenters followed Boyle's caution in evaluating the philosophical arguments. Boyle returned triumphantly to the disputes by adapting a siphon into a J tube, in which a short sealed arm containing a quantity of air could be balanced by a column of liquid in the long arm. The apparatus was troublesome, since accurate data required a uniform bore in the tube, a seal that would hold under pressure, and arms so long that they were subject to breakage. Yet in this new apparatus, volume of air (as opposed to density of air, the older concept) could be measured against pressure (as gauged by the height of the balancing liquid column). With this apparatus, as well as another piece of converted apparatus, Boyle gradually accumulated the data necessary to advance the conjecture bearing his name, although his perception in his philosophical milieu differs somewhat from our perception.[57] This example shows several interesting features, as Webster points out. Originally, theoretical dispute (albeit partly philosophical) controlled the direction of experimental research. Group enterprise and communication were essential to the development of suitable equipment and experimental technique. Boyle's discovery, although anticipated by others, was grounded in his case in superior apparatus and suspension of belief about theoretical attitudes. Boyle's law is an experimental discovery of considerable magnitude, since it is perhaps the first of the modern functional equations, in which reciprocal causation occurs between related variables, a change in either resulting in a calculable change in the other.

The next example is a study of the rise of valence theory in chemistry as a reshuffling of the themata of the radical and type theories that had preceded it.[58] This period of chemistry lasted from approximately 1830 to 1860. It begins with the discovery of the voltaic pile, an instrument that could be used to decompose chemical compounds. The electrical nature of the decomposition led to the proposal that every compound substance was composed of an electrically positive and an electrically negative part. These parts were the so-called radicals of radical theory. Radical theory was able to decompose all sorts of inorganic compounds, but there were anomalies. On the one hand, the same radical (chlorine, for example) might be a positive radical in one compound and a negative radical in another. The theory had to be sufficiently complicated to deal with these data. Further, organic compounds composed of large numbers of the same elements couldn't be conveniently divided into radicals, and sometimes couldn't be decomposed at all. Here we have the discovery of new apparatus giving rise to a new and quite general chemical theory. As electrochemical explanation failed to provide an explanation of the structure of compounds in all cases, type theory began to develop in which chemical reactions were used to organize compounds, those compounds participating in analogous reactions being regarded as of related type. The chemical character of a compound is thus to be related to the number and arrangement of its constituent atoms. Water, which binds an H (hydrogen) to an O (oxygen) and an H, is of a similar type as ethanol, which binds a C_2H_5 to an O and an H.

Clearly, both radical theory and type theory were able to organize and explain a great many isolated chemical facts, and adherents of both theories were successful in guessing the number and nature of the atoms constituting certain important compounds. The gap between data and theory here is enormous, since most chemists assumed in the nineteenth century that the internal structure of molecules could never be directly observed, partly because of the apparent limits of microscopy. Both radical and type theory were progressive at this point, producing detailed experimental data and fitting them into conceptual schemes. It was, however, increasingly evident that both the number and arrangement of atoms, as well as their electrochemical nature, were involved in chemical structure, and that this complex fact had to be incorporated into the study of organic compounds. In 1858 Kekule effected a reconciliation of the two theories. Kekule began to take formulae as pictures of the structure of compounds, and not just as a record of the input and output of reactions. Radicals remained as groups of atoms not affected by certain reactions, and as structures (types) that could play a role in other reactions. His resulting formulae were a huge advance, and his recognition of carbon-carbon bonding represented a huge step toward an understanding of organic chemistry. Like the development of quantum theory mentioned earlier, we have in this case the stimulus of new experimental data leading to rival theories that are resolved in a successor theory. Unlike the case of Boyle's law, the general direction here is from experimental data to theoretical accommodation of large scope, a pattern frequent in historical chemistry. An interesting feature of this example is that neither radical nor type theory was falsified or degenerating when each was overtaken by valence theory, nor was valence theory the result of revolution in that it resulted from the crossbreeding of prior theoretical ideas along with the somewhat mutant idea of carbon-carbon bonding, which may have been suggested partly by the discovery of gas molecules to be found in an uncombined natural state.

The third example to be considered here is the overthrow of parity conservation.[59] This example from twentieth-century physics is interesting because it illustrates the distinction in role between the experimental physicist and the theoretical physicist. Perhaps because of the severity of this functional split, besides the experimental and theoretical papers of modern physics, one can identify phenomenological papers that build mathematical models of selected experimental data, so that data and theory are more easily related and accommodated to each other.[60] The history of particle physics of concern here extends from 1950 to 1970. At the start of this period, experimental and theoretical activity were relatively independent. In 1952 and 1953 a puzzle emerged from experimental data that was an extension of previous experimental design. The decay of a certain meson (a positively charged particle) seemed to lead to two possible outcomes, a result suggesting that there might, in fact, be two different particles. While experimental evidence accumulated that there was only one particle in terms of mass and lifetime, a phenomenological argument was produced showing that if parity was preserved, the two decay modes meant that there were two particles. By 1954, theorists had become embroiled in the question of whether there were or were not two different particles. In 1956, Lee and Yang produced a theoretical argument questioning the existence of parity conservation, and pointing out that its conservation in the relevant context was not well supported by evidence. They proposed several experiments to settle the matter, and this stimulus provoked new experimental activity that demonstrated parity nonconservation within a few months. This, in turn, triggered intense theoretical activity designed to explain parity nonconservation. The *V-A* theory of weak interactions was proposed by Feynmann and Gell-Mann in 1957, a theory rich in deductive consequences subject to experimental test, and this challenge was accepted by experimentalists. Partly because of the separation of roles in physics, this example shows how complicated the interaction of theory and experiment can be over a short period of time. There is at first independence, followed by an experimental result that provokes new theory, after which theory leads its experimental testing. The relationship could hardly be more complicated, since all of these events take place in about a five-year period within one specialty within physics.

In addition to the criticisms that have already been offered of the nondialectical accounts of scientific progress in earlier discussion, it is clear that these accounts cannot survive a budget of detailed historical examples, or even the three examples that have just been discussed. Gay remarks directly that her account of chemical history is incompatible with Lakatos's methodology of research programs in that radical theory and type theory were both progressive when they were overtaken by valence theory, and White, Sullivan, and Barboni remark that their account of particle physics history is also incompatible with Lakatos's methodology in that particle physics seems intuitively to have been progressive even when its excess empirical content was declining.[61] In the work of Popper, Kuhn, and Lakatos, there is a tendency to suppose that theory must control the direction of experiment, and that history revolves around the relationship of theoretical structure to data. All three examples we have looked at exhibit periods of scientific investigation in which experiment is directing theory, and situations in which the significance of data is determined to be at odds with the prevailing theoretical climate. If theory and experiment can be separated as the moments of scientific progress for analytical purposes, the extant philosophies of science have confined themselves largely to the development of theory to the exclusion of experiment. Even Kuhn, whose descriptive account seems most easily to conform to scientific practice, has missed this division of labor, and seems not to have dealt with the fact that only some anomalies will be seen to be significant to theory, and to cause its reshuffling. The dialectical account attempts to redress this imbalance, and to relate theoretical evolution to instrumental development and the perceived significance of data. It could be embarrassed by an adequate methodology constructed along nondialectical lines, but none exists, and none seems to be in sight when the complexity of detail as a result of the burgeoning history of science is confronted. This particular dialectical account would also be embarrassed if better history indicated that the direction of evolutionary development and the extinction of theory were not clarified by apposite studies of instruments and experimental design, and in the process rendered less dependent on postulated genius. Except for this constraint, the dialectical view we have developed places almost no theoretical restraints on scientific history, and encourages involvement with the details of scientific situations without preconception about the role of theory or experiment in influencing the other.

APPENDIX

THE HUMAN SCIENCES

Our general discussion of methodology has observed the tacit convention of citing scientific examples primarily from the natural sciences, on the grounds that these examples are noncontroversially examples of science. But general reflections on methodology, if they have any empirical or theoretical content at all, should be expected to have some consequences for the disputes about methodology in the natural and the human sciences.[1] In this appendix, the consequences of our reflections for these disputes will be traced in broad outline.[2] We shall accept the intuition that there seems to be some difference (typically) between the natural sciences and the human sciences, but we shall not find it in a methodological distinction consistently with views expressed here about the irrelevance of methodology.

Both the human sciences and the natural sciences can be regarded as having theoretical and factual levels, and we would anticipate the same free play of constraints between these levels over time in both areas. To begin with, we can consider the views that the human sciences are separated from the natural sciences either because the objects they study are different, or because there is a methodological difference in the relationship of theory to fact.

Some writers have assumed that the nature of the human being, because of free will or other properties, means that the explanatory structure of the natural sciences will be simpler than, or different from, that of the social sciences. But insofar as free will implies unpredictability, it is clear that systems in both the natural sciences and the social sciences are unpredictable, and that observational exchange of information with both physical and living systems as currently defined may preclude prediction. This is true even of deterministic systems in the single case where measurement may have an incalculable impact on the system, as in some areas of quantum physics. Further, human beings can be studied in biology, or in economics, from a standpoint that seems to be that of the natural sciences. There is a tendency for subject classes to be theoretically homogeneous in the natural sciences, and heterogeneous in the human sciences, but this is not an exclusive division, and the tendency is not verified in many special disciplines. The fact that human beings may not be understood unless their entire social setting and even its history are considered can be met by the observation that cosmology involves the study of large interconnected systems and may, if relativity theory and quantum physics are theoretically integrated, have a virtually unlimited scope of interest. And although history plays an important role in the human sciences, it does so as well in cosmology and biological evolution. The attempt to find a general distinction between the natural sciences and the human sciences in terms of differences in the objects of study seems merely to repeat a prejudice, no matter that each science may have unique subject matter.[3] A general explanatory property for the presumed division has not been proposed.

General methodological differences are usually traced to differences in the relationship between theory and fact. For example, it has been claimed that this relationship is one of *explanation* in the natural sciences, but one of *understanding* in the physical sciences.[4] This cannot work as a general distinction, since both sciences depend on the notion of interpreting or understanding data text on our view, so that understanding is not limited to the human sciences. At the same time, explanation in terms of general laws is clearly sought by many theoreticians in Marxism, psychoanalysis, and other disciplines clearly to be associated with the human sciences. The current failure to have achieved explanatory insights in the human sciences cannot, without further argument, be taken as an intrinsic failing. Desired but unavailable explanation of data is a commonplace in the physical sciences.

We are not born scientists, but become scientists through the acculturation process described above. In this process, our understanding seems to engage explanation in the human and natural sciences typically in a different way. As human beings, reflecting on our own society, we tend to feel that we understand the general significance of various facets of human motivation. When we encounter an alien society, we attempt translation of their understanding of their actions into ours (which may produce an enlargement of our understanding) before we can comment on their actions. In both cases, we use this understanding to guide the process of gathering data text. If the data text is unexpected, we may well first expect that it has been gathered incorrectly. Here is a simple example. A personal but informal survey of the relevant institutions may convince us that one set of schools is better than another, a conviction that may not square with our first attempts to establish this fact quantitatively. Rather than accepting the first data as a refutation of our views, we may redesign the experiment to see if we can get the desired results. Now theory *is* constrained by data here, for if our conviction is wrong, we will never be able to locate a plausible quantitative method to prove our opinion correct, but people with differing intuitions may spend considerable time in disagreement while they redesign experiments. This is not definitionally characteristic of the human sciences, since a similar guiding of data can be observed in medicine and biology, and is undoubtedly active in new fields of physics, especially historically. In summary, when we acquire expertise in the human sciences, we are likely to find that we understand a theoretical level best intuitively, and we may allow this understanding to guide the search for plausible experimental or observational data.

Nearly the opposite may happen in the natural sciences, especially after an experimental tradition has embodied previous theoretical insights into the instrumentarium. We have no understanding of the objects of investigation, unless they are easily accessible to the human senses directly, save through the data text and previous theorizing, but we may feel that we understand how our instruments work well enough to accept the data text they produce as requiring interpretation and explanation. Thus, unexpected data text from instruments assumed to be reliable can seem to refute prior theorizing at one shot, and can call forth a variety of new theorizing constrained only by existent data. Again, this is not characteristic only of the natural sciences, since new values for social or economic indicators that have proven reliable in the past may have a catastrophic effect on theoretical predictions and call for new ideas.

Logically, understanding may enter the natural sciences or the human sciences at either the theoretical or the factual level, and we have examples of all of the possibilities. But there is a tendency for prescientific understanding, and then developed scientific understanding, to enter the human sciences at the theoretical level and the natural sciences at the level of data text. This tendency, rather than a methodological distinction, accounts for the salient facts often taken as evidence of a methodological split. In the natural sciences, we typically advance most quickly by making new instruments, or by applying established instruments and techniques in new areas, and then constructing experiments in a logical space that becomes intelligible as theory advances. We may wonder whether we've really measured "anger" or "alienation" in a human sciences experiment, and what kind of an object the "quark" could be that we've postulated to explain an odd lot of data. This difference is explainable by the typical point of entry of understanding into explanation in the natural sciences and the human sciences, a reflection that eschews any intrinsic methodological difference between them.

If low prior understanding calls for mathematical treatment, we can expect that data will be treated mathematically in the social sciences, and theory mathematically in the natural sciences. Although data are often treated mathematically as well in the natural sciences, they can often be summarized satisfactorily informally, or the details left out in a summarizing curve, while just the opposite frequently occurs in the social sciences, where the detailed data are worked up with elaborate statistical treatment to determine whether they are consistent with a theory that can be satisfactorily informally summarized. The social sciences are here taken to represent the most mathematical of the human sciences. Our account helps to explain why data from a new instrument often cause a theoretical explosion in the natural sciences, while new statistical techniques can leave data uncertainty unresolved in the human sciences. The human sciences, on the other hand, have often been most stimulated by the relatively discursive treatment of intuitively plausible ideas about people, as in the works of Weber, Durkheim, Freud, and Marx, which scientists still attempt to test against experimental observation. Or to put this another way, in the natural sciences an experiment is frequently seen to be important (Planck on blackbody radiation) long before it receives a settled interpretation, while in the human sciences theory can frequently be seen to be important (those just mentioned) long before precise confirming data are located. Theory can remain "interesting" in the natural sciences even though it is known to be wrong, but not data, and the situation in the human sciences is more nearly the reverse. As a corollary, if a physicist produces wrong data, it's a serious matter, since it can deflect the progress of physics, whereas data can't be wrong so much as misleading or irrelevant in the human sciences if taken seriously. Unexpected data in the physical sciences are thus examined and carefully repeated before publication is considered. The account of the tendency in engaging understanding thus tallies nicely with a great deal of the informal phenomenology regarding the difference between the natural and human sciences.

There is, however, an important historical difference in the instrumentaria of the natural sciences and the human sciences. The instruments of the natural sciences have in many cases replaced the organs of human sense perception. In this way, the succession of instruments defines a direction of progress. New instruments see farther, hear more distinctly, than their predecessors, and they may integrate the settled theories used to explain older data.[5] The newer instruments trade on the success of the older ones, and because the instruments are distinct, the data domains they define are distinct also. Explanation seems to reach completion because the instruments define their boundaries, and then theory can exhaust the content sufficiently to reach almost certain predictive success. But in the human sciences, the same domain tends to be divided up by thought, and no new instruments have appeared to define a succession of data domains. It may seem that two phenomena have no functional relationship, like the price of potatoes in Idaho and the number of Russian missiles, but thought can always provide the possibility of a connection. Thus in thinking about humans and society, it is hard to draw a line inside the entire domain and settle for logically possible explanations of the resulting data. Human sciences have had essentially one empirical method for studying individuals in society, that of the questionnaire, and sophistication in its form hasn't found a way of limiting the domain of theory. Because of this, the empirical domain of the human sciences has remained nearly constant except for extension in the same conceptual space for over one hundred years, in spite of the fact that no intrinsic methodological difference can single it out as intrinsically immature or different from physics. What the human sciences require for more dramatic progress is not simply more data (of the same kind), as many empiricists have stated, but new instrumentation for obtaining data, or reasonable theoretical restrictions of data domain so that more exhaustive explanatory possibilities can be tried.[6] It has not been the absence of theory, but the constraining effect of intuitive theory on the interpretation of data that has produced the current feeling of vagueness about human science explanations. Not too little theory but too much, particularly when its hold on data is so very difficult to break.

One can attempt to make the human sciences like the natural sciences, by looking for restricted data domains, but full sensitivity to the phenomena under investigation seems to constantly threaten to push one back against the full complexity of the human in society. In fact *natural science* is probably an elastic category, one that defeats any neat separation of the sciences, as Freud seems to have thought.[7] A psychology that leaves the natural science haven of stimulus-response conditioning for slips of the tongue seems inevitably drawn into the horror of the data complexity about dreams. For individual slips of the tongue, plausible scientific explanations can be attempted, and an enlarged psychology is close to normal physical science. Dream material loses its anchor in the publicly observable event, and threatens to demand a new method. Freud may have thought that if the method adequate to the phenomena could be called forth, it would be seen to be a form of natural scientific explanation, but the textual evidence is uncertain. We will attempt to solidify our conception that general methodology is irrelevant to specific differences between human sciences and natural sciences by looking at some specific examples from economics and sociology.

Both the existence of the levels of theory and fact in the human sciences as well as our considerations of the entry level of understanding into the human sciences are confirmed by the historical relationship of economics and sociology. These disciplines are often in frequent contact in the study of economic man in a social setting, and we will ignore here their troubled relationships to the psychological study of the individual.[8] The economic tradition, neoclassical and Marxist, attempts to enter the explanatory framework of the human sciences in the manner of the natural sciences, but it has been frozen in development by the lack of a suitable instrumental means of distinguishing data domains.[9] A long history of bitter argument between the neoclassical and Marxist traditions over the possible ideological status of the former in defense of the capitalistic status quo and its lack of historical insight is mitigated in areas like economic planning, where the nature of the data domain of current goods and services and short-term projections over this domain have forced a narrowing of formal differences in spite of philosophical opposition. What they have jointly produced is the most human of the natural sciences, or the most natural of the human sciences, but at any rate a discipline somewhat on the border of the intuitive distinction between the domains.

Opposed to mathematical economics, which tends to constrict to an apologetics based on appeals to some conception of theorizing in the natural sciences, is an intuitive sociology that refuses to see the economical as a natural subdomain of human science subject matter. These thinkers, historicists and holistic Marxists, find any thought of economic man to engage in principle the whole contradictory web of society, producing a domain of discussion whose complexity seems to call for a special hermeneutical or reflective method. Their entry to the human sciences is often through philosophical reflection, and the dominance of theory in their approach causes data to become a completely secondary matter. From this perspective, mathematical economics is a peripheral subject of no real theoretical interest. There is thus a tension between those for whom economic man is a legitimate abstraction and those for whom total social structure is the only legitimate subject matter of the human sciences, a subject matter that can only be illuminated, but not exhausted, in special disciplines.

Our position here is not that general methodology can resolve this dispute, but that this dispute is what should be expected in the light of the entry-level problem of understanding for the human sciences. No intrinsic reason why an instrumentarium cannot be developed for the human sciences has been found, so that this situation is the result of investment in the entry positions and a history of pushing compromise ideas into the other camp. We will examine this situation in the context of an important dispute between Adorno and Popper, which will have no purely methodological resolution. The difficulty of finding distinct data domains inside the human sciences is reflected in the instability of attempts like Max Weber's to ground human scientific investigation in the middle ground of ideal types, or Habermas's attempt to mediate between natural science and human science with a scheme of communicative competence.[10] Both the holists and the mathematical economists have seen such attempts at mediation as complete failures, indicated by the fact that the mediators are not welcomed, but placed in the camp of the enemy. One might say that the lack of an appropriate data domain of intermediate scope means that these theoretical attempts are free floating, and can be pushed away by vocal adherents of each of the oppositional tendencies provided by our analysis. Although we cannot resolve the dispute in terms of the analysis provided here, the structure of the dispute as we shall briefly consider it tends to support the methodological analysis that has been provided.

Neoclassical economics is similar to physics in that its theoretical units are all alike, economic human beings who are maximizing expected profit or expected utility. This means that by assessing the objective situation in which an individual finds himself or herself, and assuming some form of rationality postulate, we can predict that the individual will act to maximize utility, the firm to maximize profit, and so forth.[11] It may then be said that such optimizing individuals frequently have only one option open to them, or at least very limited options. In other words, if the situation is described as one of perfect competition involving goods that can be arbitrarily divided, and so on, and price and cost curves are well defined, and we assume a producer is out to maximize profits, understands the situation as described, and is rational (acts appropriately), we can predict the output level the producer will seek. In such situations, there may be only one such point on the cost curve representing optimization. We expect the producer to be near this point when we investigate empirically. This turns the decision maker into a cipher or a computer, and the logic of the decision lies fully in the situation. With these constraints, explanation of rational action can be more or less fitted into the standard explanatory framework of the natural sciences.

Neoclassical predictions have frequently failed in practice, a point taken by institutionalists, Marxists, holists, and other critics to indicate a failure in modeling.[12] The reply of neoclassicals is methodologically sound. Falsity and failure have not proven devastating in natural science provided that predictions are close enough to reality to be helpful. Further, where optimization cannot be traced to a single empirical variable in practice, the theory can be revised by a function maximizing at least two factors, a complication that parallels the adjustment of natural science theory to data. Kornai, for example, has proposed splitting the analysis of the firm into a part concerned with the production of real goods and services and a control part that processes information. In this view, the dubious neoclassical assumption that information spreads instantaneously can be combined with a theory of information flow that would allow the behavior of firms to fit empirical data more closely.[13] Other attempts have tried the addition of an explicit psychological component to handle motivational complexity.[14] But perhaps the most interesting attempt to increase the sophistication of neoclassical theory is the addition of class theory as a determinant of action, a move that brings mathematical Marxism into potential methodological congruence with neoclassical theory.[15]

Some critics of neoclassical theories have pointed out that microeconomics and macroeconomics in their current form can't be easily reconciled, that there are divergencies between neoclassical economists on specific problems in microeconomic theory, and so forth. At this level of abstraction, such observations are not likely to refute neoclassical outlooks, although they may stimulate the rate of research and development. The objections cited are of a piece with the attitudes underlying the unified science hypothesis. There is no *a priori* reason to suppose that all of economics would have to be integrated into one logically consistent economic theory before it was methodologically respectable. As we have seen, the reputation of physics as a paradigm science has not been tarnished by the fact that its theories are hardly integrated into one logically consistent theory.

The relationship between Marxist and neoclassical economic theories has been illuminated in recent years as the result of a controversy about the internal consistency of neoclassical theories based on the work of Sraffa.[16] In fact Sraffa's contention that physical production plus the wage rate determines value quantities and profits and prices without there being a causal connection between value quantities on the one hand and the latter two on the other is a powerful criticism of a number of economic theories, neoclassical as well as Marxist. As against neoclassical doctrine, this critique says that the physical description of productive mechanisms sets a range of possible rates of profit, with the actual rate of profit being set by the actual wage rate. The actual wage rate, however, is an outcome of class struggle, which is not a concept in the neoclassical domain. Neoclassicism is therefore inconsistent in the sense that its set of concepts is not sufficient to explain what it wishes to explain. (Economists use the notion of inconsistency in this way.) It can be argued in return that this criticism applies primarily to the version of neoclassical theory used for simplified exposition in which individuals (their preferences and endowments) are analytically separated from firms and their technology. In general equilibrium theory, however, one describes a system in which factor prices (including labor prices) are determined partly by individual preferences for leisure rather than for work, and individual preferences for immediate rather than delayed consumption. General equilibrium theory is a theory in which everything determines everything else, so to speak, and in which initial endowments and preferences can't be separated from technological possibilities. The theory is then perhaps not inconsistent, but it can't explain how these endowments and preferences come into being. One then confronts directly the question of whether historical changes of certain kinds are an appropriate object of economic inquiry. Because of the technical problems involved, as well as the remote empirical consequences of the theory, we cannot pursue this further here except to note that the charge of inconsistency doesn't seem to terminate the tenability of neoclassical theory.

We shall not develop formal Marxism here except to say that in newer, more mathematical versions, it can enter into explicit controversy with neoclassical theories. The resistance to class theory, in view of the fact that the concept of class is methodologically merely a theoretical posit, can arise because the theoretical use of class notions threatens a charge of complete theoretical blindness for neoclassical theorists, who have simply failed to *notice* a fundamental variable if Marxist class theory is true in some form. Neoclassical economists have also found various historical prognoses involved with class theory to be empirically dubious. The dialogue produced by Sraffa's critique is sufficient to indicate that these theories are roughly in the same methodological space, although it has taken this work to ascertain that fact with relative assurance. Holistic alternatives to economic theories relying for methodological insight on the natural sciences will now be considered.

The criticism of various historical schools of economics is that the method of economic abstraction produces economic theories that do not fit social reality, and perhaps cannot fit social reality. Insofar as the historical school simply attacks the fit of abstract theories to reality, it cannot legitimately draw the conclusion that the axiomatic method is wrong. It can only show that the axioms have been poorly chosen. To argue that the laws of physics are general, but the laws of economic development are peculiar to countries, and are therefore historical, so that economic laws are not equivalent to physical laws in terms of generality does not show a methodological difference between economics and physics. Rather than review a lot of bad arguments, we will present the following overview. The historical school began in the nineteenth century by simply eschewing abstract theory, and by confining itself to a close description of economic phenomena and their history from which it was attempted to induce "trends" on the basis of which economic recommendations could be made to prevent specifically anticipated human misery. In this procedure theory is not separate from fact, but both are seen as aspects of the particular historical data. As time went by, the abstract theories of neoclassical economics began to become more and more descriptively adequate with the development of econometrics. The response of the historical school was then to move from a detailed study of appearances to the view that economics must deal ultimately with a "hidden reality." Then the mathematical side, no matter how well it fits theories to the restricted reality that it can measure, must be wrong about the nature of man in a total societal setting. This modern form of historicism must depend on a special method, dialectical or hermeneutical, to allow the reality hidden from mathematical abstraction to come into view. The first clear point at which a special method for economics and sociology has seemed to many to appear is with the methodology of *verstehen* and *ideal types* discussed by Weber.

The value dispute in whose context Weber attempted to synthesize natural science methodology and historicism arose in a quite definite context involving the *Verein für Sozialpolitik* (a professional organization), and the question of how, as a professional body, its internal theoretical disputes and the investigations of its members should be conducted. Again, the origins of economics and sociology as we know them are involved in the dispute, but the focus of the problem is how technical problems are to be approached. Weber's contribution to the 1909 meeting of the *Verein*, as well as his call in his lecture for a value-free sociology as the legitimate pursuit for members of the *Verein*, is the starting point for a long (and still continuing) discussion of the place and meaning of values for sociological and economic theory that is known as the value dispute. As Menger had earlier attempted to separate economic theory, economic history, and economic policy, Weber separates what he calls social *science* from social politics. Social science is to be value free, although it has a special method dependent on *verstehen* and on the consideration of ideal types. Because of the complicated, eclectic views of Weber, he has for various reasons been normally primarily associated with the positivistic tradition in the German view of sociological history, but with the historical tradition in the American and English views of sociological history. Weber was concerned to discuss subjectivism and rationality in the context of the historical school. In Weber's opinion, the historical school had assumed that because human action was "subjective" in nature, it would possess irrational or arational characteristics. Weber thinks that "free" subjective action can be analyzed in terms of chains of rationality, chains linking motives or purposes to the means of satisfying them. "Irrational" behavior is insane behavior, or the behavior of people whose motives or purposes we do not understand. Once we have understood motives and purposes (through a process of *verstehen*), we will expect rational agents to act in accordance with certain laws of behavior linking these motives and purposes to the possible means of satisfying them. The use of *verstehen* is not simply to suggest hypotheses that can then be understood and validated independently of *verstehen* in the context of justification, as the functionalist's or positivist's tame version of *verstehen* would suggest.[17] *Verstehen* is used to develop hypotheses that, when understood, can be put to an empirical test only in terms of this understanding. This was Weber's concession of the historical side.

On the other hand, Weber did not see these hypotheses as subject to the test of crucial experiments prevalent in the area of the natural sciences.[18] Weber maintained that the propositions of economic and sociological theory involve concept constructions peculiar to the social sciences. As theoretical propositions they would be "exact," as are the propositions of the natural sciences, but they present a model of idealized behavior rather than a law. The *ideal types* involved in these propositions are an enhancement or sharpening of reality, as are the propositions of the natural sciences. Humans plan actions using these ideal concepts, but their planning may go astray through biased or wrong calculation, or through failure to anticipate some real world interference. The propositions underlying rational action are not hypotheses under which real actions are subsumed, nor do they stand for statistical averages of real actions. Rather they outline the course that rational human action would take in certain fixed circumstances if it were directed toward an unambiguous goal and if it were not disturbed by error and other irrational features. Weber thus thinks that the social sciences involve a separate methodological scheme, but that this scheme can be put into a fixed relationship with the methodological scheme of the natural sciences and brought under precise methodological control. But without an empirically attached data domain, this conception can slide either into natural science or into philosophical holism when other writers fail to see it as an attempt to locate a unique subject matter for the human sciences.

It's pretty clear that if the impact of *verstehen* is watered down, it is possible to read Weber as a neoclassical apologist and even as a Popperian, as many German scholars have done. As a result of the same fact, one can read Weber as a neoclassical apologist and primitive Popperian who is largely intelligible but has a slight eccentricity regarding the method of *verstehen*, as American and English scholars have done in making Weber a representative of the historical school.[19] Weber quite clearly had no idea that his methods would be applied to trivial subject matter. He had in mind that sociology should investigate the general cultural significance of social and economic structures in total human communities, as his wide-ranging studies of India, China, and Europe indicate. Weber thought the national limitations of the historical school could be transcended by his method, and that his method would result in no simple-minded reduction of history to economic forces, but a wide-ranging yet precise characterization of such cultural factors as religion.

In assessing Weber's notion of value freedom, it is essential to remember that Weber was trying to carve out a conception of economics and sociology as sciences.[20] Weber accepted a fact-value dichotomy on philosophical grounds, in which facts could be grounded in experience but values remain ungrounded. As long as sociologists and economists linked fact and value in their work, they could not achieve the sort of agreement that was characteristic of the natural sciences. This follows from the fact that value disagreement couldn't be rationally resolved within their framework. It doesn't follow from this that values won't appear in economics and sociology. Rather, given certain values, appropriate lines of action will be objectively discussable within these disciplines. In the same way that a physicist's personal feelings about physical particles should not matter in physics, an economist's feelings about, say, socialism should not figure in his or her scientific analysis of socialism—and to conceal such involvement deliberately would be the mark of a scoundrel. The sociologist or economist must therefore externalize values, express them precisely, and study their relationship to each other and to fact. In this way, these disciplines can be said to have a detached and objective subject matter like other scientific disciplines. The eclectic genius of Weber is evident in the doctrine of value neutrality. Where many historicists thought that increasing technological power would fuse values and differences of opinion into a postideological society, Weber agreed with Marxism that only an increasingly intensified class conflict could be expected, an opinion apparently confirmed by the events of the First World War. But as against the prevailing Marxism of his day, Weber held that one's thought need not be determined by class standpoint and associated set of values, but that a scientist could externalize class conflict and study it from a value-neutral point of view.

Critical theorists and some Marxists could argue that fact and value are not separable as Weber assumed, that values are not independent of rational grounding as Weber thought, and that the method of *verstehen* can only yield an analysis of the misleading appearance of conscious thought and social totality where there are class conflicts, and not an analysis of social reality.[21] The critical theorists have held all of these positions, and have hence lumped Weber with Popper (and even the positivists) as having an insufficiently reflective and self-critical social philosophy. By way of anticipation, the following remark by Habermas concerning Parson's appraisal of value freedom in Weber is crucial:

> Please allow me a final intellectual historical observation. Parsons has claimed that Max Weber's teaching is a development towards bringing about the end of ideology. Weber is said to have broken the trilemma of historicism, utilitarianism, and Marxism, and to have led the way into the free field of discussion beyond the European fronts of civil war. I envy our American colleagues their political traditions which permit such a generous and (in the best sense of the word) liberal interpretation of Max Weber. We here in Germany, who are still seeking for alibis, would only too gladly follow them. But Weber's political sociology has had a different history here. At the time of the First World War he outlined a sketch of a Caesar-like leader-democracy on the contemporary basis of national-state imperialism. This militant latter-day liberalism had consequences in the Weimar period which we, and not Weber, must answer for. . . . Viewed in the light of

> the history of influences, the decisionist element in Weber's sociology did not break the spell of ideology, but strengthened it.[22]

In the *Positivismusstreit*, or positivist dispute, there is the major discernible problem of whether or not German sociology is to be dominated by American empirical sociological methods and theory, which were clearly formed by positivist criteria of what is to count as scientific.[23] There is also the problem of whether German sociology is to be dominated by reflex of mathematical Marxism. This debate from the 1960s is of great clarifying potential for the view we are considering of the relationship between the human and natural sciences. Adorno argues in this dispute for an autonomous (German) sociology, to be erected in some way at least partly on the insights of German idealist philosophy, autonomously because this reflective idealist component would not allow it to be merged with, or controlled by, restricted empirical domains. On the other hand Popper, who is the major antagonist of Adorno in the early stages of the debate, even if not a classical positivist, is clearly not attuned to any possibility that scientific method is other than perfectly general, that is, common to all scientific disciplines and having the same force in all social settings. Thus we have in Popper's view the idea that the method of sociology and economics involves a rationality postulate distinguishing it from the method of the natural sciences, but also a view that a general methodological scheme can be worked out for all of the sciences that is valid everywhere and explicitly depends on a single model of scientific explanation. This attempt to ground German sociology on the manifestly successful practice of American sociology, as interpreted by Popperian methodology, is clearly resisted by Adorno. The title of Adorno's earliest essay in the positivist dispute, "Sociology and Empirical Research," suggests by itself that sociology and empirical research are two distinct things. The dispute between Adorno and Popper thus represents the clash of general methodology with the view that the human sciences must be coerced toward holism by their data.

We will now initiate a discussion of the positivist dispute in some detail, at least the initial confrontation between Adorno and Popper. Popper thinks that science can be demarcated from philosophy and metaphysics, that science is the only hope for the future of mankind, that science (if retained in a proper context) yields an optimistic forecast for the future of mankind, that science at a time has a structure that can be clearly articulated by philosophers, and that the relevant parameters of science (explanatory and corroborative) can be calculated. In these respects he shares important doctrine with the positivists, even though he rejected their key explications of the doctrine of confirmation. Adorno regards as positivistic any naive starting point in the accumulation of data that holds that the relevant data can be described in terms of existent unreflective categories. Some of the dispute between Adorno and Popper about positivism is thus merely terminological, and we must circumvent this if we are to come to terms with their more crucial confrontation.

Popper later complained that his remarks opening the 1961 Tübingen conference, the first round of the dispute, were not explicitly discussed by Adorno even though he presented them as twenty-seven theses and invited Adorno's explicit agreement or disagreement.[24] What this means is that Adorno did not explicitly agree or disagree with Popper's theses. Adorno could not have done this, since he had good grounds for supposing that the resulting debate would have been biased toward Popper's position as a result of his acceptance of Popper's language as the basis for discussion. On the other hand, it is not true that Adorno did not reply to any of these theses. Adorno quite explicitly repudiated many of them. Let us begin by looking at Popper's theses. The first three collectively state that we have constantly increasing knowledge but that we are also always ignorant. In the fourth and fifth theses Popper takes the origin of knowledge to lie in the selection of problems. He must, of course, take some such tack, since according to his doctrine of falsifiability, knowledge cannot originate in observation, and as conjectured it is always capable of refutation. Thesis five suggests that we can draw a distinction between significant and insignificant problems, that is, that we can tell which problems are the ones most worth discussing. Popper believes we are easily able to determine what our problems really are, so to speak. This thesis raises a direct confrontation with the thrust of critical theory. First, it lacks a reflective component. It may be that at least one source of knowledge is reflection on the problem space in an effort to determine what our problems really are, and a reflection that may involve considerable social scope. The epistemological position defended above is that significance can't be determined unambiguously at a given time, so that Popper's easy assurances here are problematic. Second, to see the state of science, including the social sciences, as a more or less historically contingent set of isolated problems that should be, and can be, solved independently on a piecemeal basis is once again to adopt a stance that seems to need tempering through reflection. Popper does not brood over the idea of attempting to make sense out of the total nature of the problem set in the way that critical theorists do. Popper is content to highlight the sort of short-term problems that are at the forefront of empirical research. Third, Popper finds success in the social sciences to be directly proportional to the "honesty," directness, and simplicity" with which problems are tackled. Again, this seems naive. As honesty is not to be confused with the intention to be honest, and directness and simplicity become more and more obscure as one contemplates what they might mean in this connection, this advice seems to come down to the idea that one roll up one's sleeves and get to work, that is, to nothing very clear at all in this context. By contrast, patience, deviousness, sophistication, reflection, and even other properties may take on a virtue of their own. Thus Adorno is staggered by what he sees as the naiveté implicit in this advice, as seems clear from his remarks:

> But the cognitive ideal of the consistent, preferably simple, mathematically elegant explanation falls down where reality itself, society, is neither consistent, nor simple, nor neutrally left to the discretion of the categorical formulation. . . . Popper objects to the cliché that knowledge passes through a series of stages from observation to the ordering, processing, and systematization of its materials. This cliché is so absurd in sociology because the latter does not have unqualified data at its disposal but only such data as are structured through the context of societal totality. To a large extent, the alleged sociological ignorance merely signifies the divergence between society as an object and traditional method.[25]

Perhaps this is sufficient to indicate that Popper and Adorno do not always talk past one another. Although there is the view that all of our knowledge is conjectural in Popper, so that any of it can become falsified, there is no suspicion that it could simply fail to fit the outlines of social reality.

We will not examine the rest of Popper's theses in detail. Many of them oppose historical determinism, on the one hand, and relativism on the other, particularly as exhibited in some forms of the sociology of knowledge. Adorno could agree largely with these criticisms, since he considers himself neither a determinist nor a relativist, and considers the sociology of knowledge to trade on the unreflective categories of the conscious mind. Popper's more positive description of his own methodology for the social sciences is of more interest. In Popper's framework, the crude unified science view of positivism is replaced by a view in which the natural and social sciences share a mode of deductive explanation, except that the social sciences must make use of a rationality postulate to make social scientific explanations applicable to idealized patterns of human behavior. Popper is thus the heir to Weber in attempting a special explanatory principle in the social sciences, while looking to the natural sciences as the arbiter of general methodology. If the rationality postulate must occur as a premise of explanation in the social sciences, however, there is little point in viewing natural and social science explanations as having a common form, since the rationality postulate is neither a theory nor a fact in the sense usually given in expositions of the deductive model of explanation. Its compatibility with neoclassical economic explanations, however, is manifest. Popper is opposed to explanation in terms of individual psychological states, or in terms of forms of life. He holds that rationality transcends a cultural setting, and that institutions or social groupings cannot have causal force in determining human action. Functionalist or historicist explanations may seem to work when people are acting in a fixed way over a period of time. But they cannot explain change in society. (On the surface, Popper has taken nearly the obverse position to that of mathematical Marxists, and the ease with which this can be done supports our contention that these positions are methodologically similar.) Only individuals can change institutions in Popper's view, so that change is always the result of some set of individual actions. But an individual's actions can be explained if and only if they are not unique, that is, if and only if they are not dependent on the unique thoughts of the individual making the decision, but can be understood as rational by others. Thus, to explain individual action, we describe choice situations and the decisions that "rational" individuals would make in these situations. This is an ideal type of explanation, variations from which are to be explained by human psychological quirks, that is, miscalculations, bias from emotion, and so on. Clearly, standard neoclassical explanations of consumer behavior fall under the rubric of such situational analysis. Popper sees the main task of the social sciences as dealing with problems that arise when the actual outcomes of action have awkward consequences:

> And this remark gives us an opportunity to formulate the *main task of the theoretical social sciences. It is to trace the unintended social repercussions of intentional human actions.* I may give a simple example. If a man wishes to buy a house in a certain district, we can safely assume that he does not wish to raise the market price of houses in that district. But the very fact that he appears on the market as a buyer will tend to raise market prices.[26]

The conservative nature of this outlook is nearly obvious. For example, a number of unfortunate experiences with urban renewal projects has brought to light the fact that people may react negatively to relocations that seem to improve their physical environment enormously. The reasons are obviously various and somewhat mysterious, but they are expressed as a felt loss of significance or identity, or a shift in the amount of "essential" versus "nonessential" transportation time.[27] This behavior can't be usefully set aside as irrational, although early planners did so. Now Popper could accommodate the idea that perhaps the situation could be described or redescribed, so that "essential" and "nonessential" transportation time could be objectively assessed and new rational explanations derived, but such examples seem to show that he downplays the role of psychological feeling too much, that his methodology contains no hint about when new descriptions of situations are required to preserve rationality, and that his methodology is consequently conservative in some root sense. Among other problems, the mere tracing of unintended consequences leaves the entire question of what to do about them to decisional choice that must be made outside of science.

These remarks do not refute Popper. He avoids the pitfalls of naked empiricism, argues against the reduction of the sciences to physics, is sensitive to various problems of the relationship of theory to observation and of fact to value. His revolutionary doctrine of fallibilism for scientific thinking is coupled, however, with his doctrine of piecemeal social engineering, and here, I believe, is where the root issue between Popper and critical theory arises. Popper believes that European liberal democracies represent the best and least repressive historical form of human society, and the only form of society that is, in the long run, compatible with the scientific tradition that he admires. By keeping the context alive, one can keep science alive, and use its discoveries to solve human problems at a rate sufficient to have a chance of reducing human misery. At the same time, he has no real point of leverage against those who don't accept his perception of European history, those who see humanitarianism within Europe and America purchased at the cost of barbarism elsewhere, those who find scientific knowledge currently an artificial human product geared to the manipulation of people and things, and those who see a sinister form of domination backed by scientific manipulation as the immediate form of European and American culture. These sharply divergent attitudes are, in fact, held by critical theorists, and this helps to explain the mutual hostility that one can feel between Popper and the critical theorists. Popper's theory of the social sciences, in placing choice of values and resultant change in the category of decision, has no resources for dealing with sharply divergent attitudes.

Adorno is simply opposed to methodology in Popper's sense of the term. Methodology is a theory of how the scientist ought to conduct his or her work, and as such its mode of operation is to endeavor to make science conform to its theoretical precepts. But this, for Adorno, is backward. No theory (and no concept) of discursive scientific language can ever fit reality, in this case social reality, exactly. Therefore, reality must be given ontological and epistemological priority in reflection, and as we come to uncover reality through a dialectical process, we must simply adjust method and theory to fit what is uncovered.[28] The idea that there is a general methodology of science wrongly idealizes science, and makes it seem a flawless route to an assurance of understanding reality, which it is not. The point of view involved can only aid science to be a tool of domination in the wrong hands. Popper's methodology constantly places reality on a Procrustean bed of falsifiability. Popper finds the wrong measure of science because he overvalues philosophy, which has always tended to give the primacy to theory. The holistic contrast thus depends on a data domain, but one that is as broad as possible within our current conceptual resources.

This general assessment of Popper is to be found at the end of the first paragraph of Adorno's reply to Popper, where Adorno says that he is interested in the concrete mode of procedure of sociology, and not in the deductive structure of the cultural presentation of its results, as the logic of sociology.[29] Adorno makes no distinction between discovery and prediction in that he believes discovery to have a logic in which the object of study is allowed to reveal itself. As a result, a good theory can be construed as a description of a "hidden" reality, but not a perfect description of that reality. Adorno makes it clear immediately that he takes Popper to have dealt with only part of sociology.

Adorno can accept the verbal idea that sociology is based on problems, expressed in Popper's early theses, but he would not, of course, recognize the same problems as the focus of development. The primacy of the object, in the case of sociology a contradictory society, is crucial.[30] Although contradictory, society is determinable. What Adorno means becomes clear in the course of his essay. Current society is a contradictory mixture of modes of production, psychological types, traditions, that can only be elucidated by a historical understanding, and cannot be adequately understood in a consistent theory. A dialectical or critical theory cannot be broken down into subcomponents. In other words, to simply study the family in modern society in terms of its statistical appearance is pointless, because the family is what it is in modern society because of the interaction of family possibilities with the rest of society. In other societies, the family is a different institution, and understanding the current family involves these possibilities and their historical link to the present. Popper's methodology supposes that you can successfully theorize society into logically independent subunits, but Adorno's view is that nothing approximating the truth could come from this approach. This point would have to be thrashed out through specific lines of research, and a great deal of Adorno's output elsewhere consists in showing the inadequacy of sociological research that doesn't see the whole society reflected in the cultural object of immediate study:

> There are sociological theorems which, as insight into the mechanisms of society which operate behind the facade, in principle, even for societal reasons, contradict appearances to such an extent that they cannot be adequately criticized through the latter. Criticism of them is incumbent upon systematic theory, upon further reflection but not, for instance, upon the confrontation with protocol statements. . . . If, in the last analysis, one does not wish to confuse sociology with natural-scientific models, then the concept of the experiment must also extend to the thought which, satiated with the force of experience, is projected beyond the latter in order to comprehend it.[31]

Adorno does not believe that the experimental method can simply be imported into sociology, since the total society cannot appear in the experimental setup. The society must be thought through if the experiment is to have any coherent meaning. Although Adorno agrees with Popper on the verbal formulation that the critical method is essential to science, he means by the critical method working through the apparent phenomena to see a reality not covered by theory, while Popper tends to believe that the phenomena themselves can be used to criticize theory. Adorno cannot limit criticism to sociological statements, since criticism must be of the society that ultimately is the object of sociological inquiry. Criticism cannot have logical consistency as a goal, since logically consistent theories and data may not be accurate representations of society. Adorno's methodological argument is that society cannot be comprehended by generalizing from a study of its parts in the manner of the natural sciences:

> In sociology one cannot progress to the same degree from partial assertions about societal states of affairs to their general, even if restricted, validity, as one was accustomed to infer the charac-

> teristics of lead in general from the observation of the characteristics of one piece of lead.[32]

A random sample of lead will have the same properties as other samples of lead, that is, as lead in general has. Lead is a homogeneous class. But the subjective reactions of people or individual institutions will not have the same properties as society in general. A unique society is the focus of sociological attention. Thus we cannot ascend from individual cases to society, except in a trivial way. Society is already present in the individual cases. Opinion research, by accepting the ideological self-consciousness of individuals that they are independent of one another and of society, makes a fundamental error:

> For the findings of what is called—not without good reason—'opinion research' Hegel's formulation in his *Philosophy of Right* concerning public opinion is generally valid: it deserves to be respected and despised in equal measure. It must be respected since even ideologies, necessary false consciousness, are part of social reality with which anyone who wishes to recognize the latter must be acquainted. But it must be despised since its claim to truth must be criticized. Empirical social research itself becomes ideology as soon as it posits public opinion as being absolute.[33]

This position avoids any lazy call for more data or more theorizing, but it eschews methodology in general as offering no clear guidelines for development, since it is caught in the picture of natural science experimentation. Its proposal that significant data be sought within the context of a coherent view about social reality, however, is quite consistent with a dialectical view of scientific development, and of a piece with the dialectical view that has been proposed here to account for the development of the physical sciences.

In the confrontation between Adorno and Popper, two positions are represented that tie a philosophical theory to a coherent critique of empirical research. Adorno felt free, for example, to constrain data on the authoritarian personality according to his reflective position on Fascism, and he was able to offer sharp criticisms of specific instances of opinion research. Popper's position represents a philosophical legitimation of the domain of data regarding the public manifestations of individual maximization of utility, an aspect of human individual behavior that can be clearly isolated in some situations and, as a consequence, is a coherent domain of investigation. Between these positions, as they have been generalized above in the discussion of entry points of understanding, economics and sociology have not located clear data domains marked out by instrumentaria to which rival theories can attempt adaptation. Theories of intermediate domains in sociology have been linked by special methods to their own data, with the result that the relationship of the theories to each other becomes unmanageably vague, and data are used primarily in discussion of the theory to which they are related at their point of origin. This confusion of sociologies is the result of attempting to extend natural science models into domains of intermediate scope.[34] On the other hand, attempts to extend the scope of reflective sociology by subdividing the domain of social totality have produced normative social philosophies divorced from the constraint of specific data. The continuation of the positivist dispute indicates this situation nicely. Habermas and Albert achieve a partial rapprochement of the conflict between Adorno and Popper, but at the expense of clear lines of criticism of specific research.[35] This development, too complicated to follow closely in a book already of inordinate length, indicates once again that science is absent where the borders of data domains are not clearly posted.

NOTES

Chapter 1. Logic and Science

1. See the discussion in Ball and Coxeter, *Mathematical Recreations and Essays*, pp. 243–250. There are many problems without known logical solutions. Further, the notion of what is a logical solution can shift with mathematical and scientific developments. Quantum physics, for example, has had an impact on the notion of the excluded middle.

2. Hume's legacy is the realization that the future cannot be *logically* inferred from the observed properties of the past. See Goodman, *Fact, Fiction and Forecast*, for a discussion of this legacy and a strategy for dealing with consistency in projection.

3. Levi has discussed this extensively, but remains optimistic about quantifying epistemic utilities. See Levi, *The Enterprise of Knowledge*.

4. Newcomb's puzzle is introduced in Nozick, "Newcomb's Problem." For a development, see discussion of the puzzle in Cargile, "Newcomb's Paradox," Levi, "Newcomb's Many Problems," Olin, "Newcomb's Problem," and Skyrms, *Causal Necessity*.

5. *Cartesian* described the philosophy of Descartes as well as the philosophical tradition based on his work, a tradition whose analyses are restricted to the content of *conscious* ideas.

6. The metaphor is given in Popper, *Objective Knowledge*, p. 121. Popper's original evolutionary metaphor involving falsifiability is later confounded with a view of verisimilitude that he uses to combat relativism. See Ackermann, *The Philosophy of Karl Popper*, for a general discussion, and Miller, "Popper's Qualitative Theory," and Tichy, "On Popper's Definition of Verisimilitude," for a discussion of verisimilitude.

7. See the extended discussion in Bellone, *A World on Paper*. I feel that I should apologize to Bellone for using his examples to support philosophical points.

8. Such claims have frequently been described as synthetic *a priori* judgments in the literature. See Hanson, *Patterns of Discovery*, discussion of footnote 1 of text, p. 104, which occurs on pp. 208–209.

9. The example of free fall is Cartesian, but not to my knowledge to be found in this form in Descartes' writings. *Suitability* as applied to this case means that the objects bear only physical relationships to each other. Otherwise a slower object might move faster out of love, for example, when coupled to an object of desire. Descartes was brilliant at making such reducing assumptions. It should be noted also that bodies do not fall parallel without constraint when dropped, but this refinement was not a factor in the relevant debate.

10. See the discussion of Husserl in Bauman, *Hermeneutics and Social Science*, pp. 111–130.

11. If theory is tied too closely to what has already been observed, it must lose its vital speculative component. For an example of the difficulties involved, see Carnap, "The Methodological Character," and the discussion in Scheffler, *The Anatomy of Inquiry*.

12. In *Logical Foundations of Probability*, Carnap was forced to the conclusion that general theories must have a probability measure of zero. He therefore reduced scientific inference in most cases to next-case probabilities. According to the features of positivism that have been singled out, an acceptance of determinism is not an essential feature of positivism. There is, however, a pronounced tendency for positivists to accept some form of determinism. In its stronger versions, determinism entails that the microstates of the universe are arbitrarily precise and that transitions between them are mediated by fixed causal processes. Indeterministic theories are now thought to be essential in many areas of science. Bastin, *Quantum Theory*, Bohm, *Causality and Chance*, and Hanson, *Patterns of Discovery*, will introduce the issues. A determinist must hold these theories to be provisional, one day to be replaced by deeper deterministic theories. If determinism were true, explanation and prediction would be symmetrical, as many positivists believe. One could regard Neo-Darwinian evolutionary theory as suspect as a result, since the theory can often only explain retrospectively adaptations that it cannot predict. We have set this discussion to one side here in order not to rule biological theory out of court on methodological grounds.

13. For examples of interesting developments, see Bressan, *A General Interpreted Modal Calculus*, and Skyrms, *Causal Necessity*.

14. The two models are presented with great clarity in Hempel, "Deductive-Nomological vs. Statistical Explanation."

15. See the discussion of explanations and explanation sketches in Hempel, *Aspects of Scientific Explanation*.

16. Hempel, *Aspects of Scientific Explanation*, is a classic exposition. The full model has various technical aspects that are simplified in our treatment without essential loss.

17. Eberle, Kaplan, and Montague, "Hempel and Oppenheim on Explanation." Consider a theory T and a fact E to be explained. Using a plausible syntactical construct of what a theoretical sentence is, the authors showed that another theory T^* could be derived from T, so that E could be explained on the Hempel model from T^* and a fact whose truth was assured by E. Since T and E can be arbitrarily chosen in the construction, T and E can seemingly be arbitrarily linked by a Hempelian explanation.

18. Proposals for restricted models are surveyed with great insight in Stegmüller, *Probleme und Resultate*.

19. This is a real incident, slightly altered to protect the living.

20. See Popper, *Objective Knowledge*, and the discussion of falsifiability in Ackermann, *The Philosophy of Karl Popper*.

21. This example is taken from Carnap, *Logical Foundations of Probability*. For more sophisticated systems and defensible Bayesian strategies, see Car-

nap and Jeffrey, *Studies in Inductive Logic*, and Rosenkrantz, *Inference, Method, and Decision*.

22. Our effort to avoid technicality runs into difficulties here. In the case of all heads, there is only one sequence. Where there are 999 heads, the one tail could appear at any point in the sequence, so there are 1,000 possibilities. Where there are 998 heads, there are obviously even more, since the pair of tails could appear next to each other, or arbitrarily far apart. The complete claim that the number of possibilities continues to rise until there are as many heads as tails is proved easily in probability theory, but perhaps this will motivate its plausibility sufficiently for our purposes.

23. A similar point holds for heads, obviously, and if there are more than 500 tails in both of two sequences with a different number of tails, that with the greatest number is most likely. Our example involves two specific cases, but no priors for the utilization of Bayes' Theorem. For the mathematical treatment that will evaluate all possible cases and calculate the relevant probabilities and confirmation functions, see Carnap, *Logical Foundations of Probability*, or the brief introduction in Ackermann, *The Philosophy of Karl Popper*, pp. 105–113.

24. Normal science is described in Kuhn, *The Structure of Scientific Revolutions*.

25. Revolutionary science, the contrast to normal science, is also developed in Kuhn, *The Structure of Scientific Revolutions*.

26. Feyerabend, *Against Method*, pp. 17–53, *passim*.

27. Feyerabend prescribes a dash of anarchism in *Against Method*.

28. For a sophisticated defense of this view, see Goodman, *Ways of Worldmaking*.

29. The unified science hypothesis is that one language can, in principle, suffice for the development of all of science. Historically, the language of choice must be adequate to physics.

30. The research reported in Knorr-Cetina, "Social and Scientific Method," shows this dramatically. A technician not expecting to see a phenomenon writes as follows in the absence of guiding theoretical considerations (p. 340):

> A murky, possibly imaginary interface appeared after about ninety minutes, separating an opaque, purplish upper layer which did not clear from a blackish lower layer. Further, the 'interface' could not be seen moving when the outlet was opened to drain off the lower fraction.

31. Ravetz, *Scientific Knowledge and Its Social Problems*, p. 116, makes an interesting comparison betwen Lavoisier's 'discovery' of oxygen and the discovery of America by Columbus that is well worth pondering.

32. That it is the weight of objects that is important to their rate of fall, and not their color or shape, is a bold insight that narrows the range of logical possibilities to be made. Experience could not by itself rule out all of the bizarre functions that logic might suggest.

Chapter 2. Social Structure in Science

1. Hanson, *The Concept of the Positron* and *Patterns of Discovery*. Hanson, influenced by Wittgenstein, attempted a context-sensitive philosophical description of scientific practice.

2. Merton, *The Sociology of Science*. Mulkay has suggested that Merton's norms provide a repertoire of moral rhetoric that scientists can use to establish their work as proper. See Mulkay's remark in Barnes and Edge, eds., *Science in Context*, p. 18.

3. Barnes and Dolby's "The Scientific Ethos" discusses changes in the norms over time.

4. Mitroff, *The Subjective Side of Science*, p. 12, provides this useful summary.

5. Watson, *The Double Helix*.

6. This problem is discussed with great insight in Ravetz, *Scientific Knowledge and Its Social Problems*.

7. Martin, *The Bias of Science*, analyzes some specific papers by scientists on the expectation of damage to the ozone layer by supersonic transport emissions. The opposing scientists chose models whose development led to predictions in line with the national priorities of their countries with respect to the transport plane, a fact that Martin thinks cannot be accidental. Another stimulating account of conflicting scientific judgments apparently based on differing social factors of institutional arrangement for making decisions is given in Gillespie, Eve, and Johnston, "Carcinogenic Risk Assessment." See also the relevant history in MacKenzie, *Statistics in Britain*.

8. Ravetz, *Scientific Knowledge and Its Social Problems*, p. 194.

9. Whitley, "Types of Science," p. 434.

10. Kuhn, *The Essential Tension* and *The Structure of Scientific Revolutions*. The influence of the latter book has been enormous, and many authors have attempted to find science through locating paradigmatic consensus and revolutions where Kuhn himself never expected his ideas to apply. See Ackermann, "Methodology and Economics," for the case of economic history. Loose characterizations of Kuhn are often used to support this process of legitimization in otherwise coherent attempts to write the history of a field. See, for example, Rosnow, *Paradigms in Transition*, p. 4:

> Kuhn defined a paradigm as consisting of scientific achievements that were sufficiently unprecedented to attract an enduring group of adherents away from competing modes of activity.

11. See Hacking's review of Kuhn's *Structure*, Kordig, *The Justification of Scientific Change*, Krüger, "Die systematische Bedeutung," Lakatos and Musgrave, eds., *Criticism and the Growth of Knowledge*, Scheffler, *Science and Subjectivity*, and Shapere, "Meaning and Scientific Change," for a survey of philosophical attitudes toward Kuhn's views.

12. Crane, "An Exploratory Study." For an argument that research communities are sociological artifacts that have no explanatory value, see Knorr-

Cetina, "Scientific Communities." See also the account in Stokes, "The Double Helix."

13. The notion of an exemplar, as well as the notion of a paradigm, has undergone some change. See Kuhn, *The Structure of Scientific Revolutions*, pp. 174–210, and Kuhn, *The Essential Tension*, pp. 293–319. An application of the notion of exemplar to the charm-color debate in high-energy physics may be found in Pickering, "Interests and Analogies."

14. Kuhn, *The Structure of Scientific Revolutions*, pp. 62–65.

15. See the important discussion of this point in Martins, "The Kuhnian 'Revolution.' "

16. Kuhn, *The Structure of Scientific Revolutions*, p. 151. Perhaps *young* should refer to freshness in a discipline rather than to chronological age. See the discussion of this point in Stokes, "The Double Helix."

17. Hull, Tessner, and Diamond, "Planck's Principle." I apologize to the authors for stealing their irresistible joke.

18. See Hanson, *Patterns of Discovery*, for a general account. It is interesting in this connection that the notion of paradigm plays no explicit role in Kuhn's own history of these developments in Kuhn, *Black-Body Theory*.

19. Krüger, "Die systematische Bedeutung," p. 218.

20. Kuhn, *The Structure of Scientific Revolutions*, pp. 205–207.

21. This is, of course, a wild simplification. Only those parts of classical physics assumed to contrast with relativity theory and quantum theory are being considered here. The so-called second revolution in physics, brought about by the expansion of classical theory into thermodynamics, radiation effects, electromagnetic theory, and statistical mechanics, and not at first subject to a fit with data, is ignored here. See the superb account in Bellone, *A World on Paper*.

22. Garfield, Malin, and Small, "Citation Data," and Griffith et al., "The Structure of Scientific Literatures."

23. Cole, Cole, and Dietrich, "Measuring the Cognitive State."

24. Mullins, "A Sociological Theory."

25. See the interesting discussion of polycentrism in Ravetz, *Scientific Knowledge and its Social Problems*, especially pp. 260–288.

26. Blume and Sinclair, "Aspects of the Structure."

27. Beyer and Stevens, "Forschungsaktivität und Produktivität."

28. Ibid.

29. Whitley, "Types of Science," p. 438.

30. Baldamus, "Relevanz und Trivialität," and Knorr, "The Nature of Scientific Consensus."

31. Elsasser, *Atom and Organism*.

32. Knorr, "The Nature of Scientific Consensus."

33. Popper, *Objective Knowledge*, especially chaps. 2 and 3.

34. Nowotny, "Controversies in Science," and Pinch, "What Does a Proof Do."

35. Bourdieu, "The Specificity of the Scientific Field."

36. Nowotny, "Controversies in Science," and Pinch, "What Does a Proof Do," p. 174.

37. Bellone, *A World on Paper*, discusses private dictionaries for some famous scientists.

38. There are many examples supporting this in scientific history. Olby, *The Path to the Double Helix*, makes interesting reading in this connection.

39. Gilbert, "The Development of Science."

40. Taylor, "Understanding in Human Science."

41. Pinch, "What Does a Proof Do," p. 181, and Whitley, "Changes in the Social and Intellectual Organization," p. 157.

42. Latour and Woolgar, *Laboratory Life*, pp. 168–174.

43. Ludwig Wittgenstein, *Philosophical Investigations* (New York: Macmillan, 1958), p. viii.

44. Lakatos, *Philosophical Papers*, Laudan, *Progress and Its Problems*, and Kuhn, *The Structure of Scientific Revolutions*, pp. 205–206.

Chapter 3. Science and Nonscience

1. Collins and Pinch, "The Construction of the Paranormal," Collins and Pinch, *The Social Construction of Extraordinary Science*, and Pinch, "Normal Explanations of the Paranormal." Collins and Pinch, *The Social Construction of Extraordinary Science*, pp. 22–24, reports how easy it is to find parapsychology a normal science when one is interacting with parapsychologists, an interesting comment on the notion of normal science.

2. Wynne, "Between Orthodoxy and Oblivion" and "C. G. Barkla and the J Phenomenon."

3. Mulkay, *The Social Process of Innovation.*

4. Clavelin, *The Natural Philosophy of Galileo*, and Feyerabend, *Against Method.*

5. Wright, "A Study in the Legitimisation of Knowledge." Hannaway, *The Chemists and the Word*, is also interesting in its explanation of how formulae in chemistry became detached from the personality of individual chemists and could be transferred to the public domain of controversy, unlike the formulae of Paracelsian physicians.

6. This picture has been presented by Carnap, for example, and can be found in *Logical Foundations of Probability.*

7. Ziman, *The Force of Knowledge*, pp. 8–36. See the discussion revolving around a newer concensus in Barnes and Edge, eds., *Science in Context*, especially pp. 147–154.

8. Hounshell, "Edison and the Pure Science Ideal."

9. Ibid., p. 612.

10. Evenson, Waggoner, and Ruttan, "Economic Benefits from Research."

11. Rosenberg, *Perspectives on Technology*, p. 269.

12. Ferguson, "The Mind's Eye," and Ivins, *Prints and Visual Communication*, pp. 51–71.

13. An effective argument that the information in visual presentations cannot be explicitly paraphrased into statements can be found in Goodman, *Languages of Art*.

14. Popper, *Objective Knowledge*, Lakatos, *Philosophical Papers*, and Kuhn, *The Structure of Scientific Revolutions*.

15. It has been possible for the adherents of various methodological schools to write the history of science from their own perspectives without residue. Apparently the facts of scientific history allow themselves to be interpreted in a variety of ways.

16. Popper has always worried that political intrusions could ruin the free criticism required for scientific progress in his view, a theme whose implications are worked out in some detail in Ravetz, *Scientific Knowledge and Its Social Problems*, as well as in Bloor, *Knowledge and Social Imagery*.

17. Hanson, *Patterns of Discovery*.

18. The problem is examined in the case of curve fitting in Ackermann, "Inductive Simplicity." For the general constraint of the past, see the approach through thematics in Holton, *Thematic Origins*, and the special case of high-energy physics in Koester, Sullivan, and White, "Theory Selection."

19. Ehrlich and Feldman, *The Race Bomb*, is a good popular presentation of the issues, but see also Urbach, "Progress and Degeneration," and the discussion of Urbach by Tizard in *BJPS* for philosophical perspective.

20. Shapin, "The Politics of Observation."

21. See the discussion in Bloor, *Knowledge and Social Imagery*, pp. 48–73.

22. Ravetz, *Scientific Knowledge and Its Social Problems*.

23. Pfetsch, "The 'Finalization' Debate," and Barnes and Edge, eds., *Science in Context*, pp. 187–195, along with papers on this topic reprinted in the same volume.

24. This disappearance has been traced by Bachelard, who argues that displaced concepts may still have an effect on creative reasoning in science on the unconscious level. Most of the important texts have not been translated, but the outline of Bachelard's thought can be gathered from LeCourt, *Marxism and Epistemology*.

25. Kuhn, *The Structure of Scientific Revolutions*, presents scientific education in terms of such an acculturation model.

26. See the discussion of Bachelard's "epistemological break" in LeCourt, *Marxism and Epistemology*.

27. Ronchi, *New Optics*.

28. Widdowson, *Explorations in Applied Linguistics*, pp. 51–61. More recently, the rhetorical style of science has been studied in comparison with that of other disciplines. The scientist may rely on consensus to argue effectively without self-assertiveness, by letting the subject matter do the talking when his or her audience is familiar with the context. See Bazerman, "What Written Knowledge Does," Yearley, "The Relationship between Epistemo-

logical and Sociological Cognitive Interests," and Yearley, "Textual Persuasion," for an analysis.

29. Conant, "The Overthrow of the Phlogiston Theory," and Toulmin, "Crucial Experiments," present the standard recognized facts about this set of experiments.

30. Bloor, *Knowledge and Social Imagery*, p. 128.

31. Historians generally recognize at present that the older narrative style of history contains questionable assumptions, but this point has not yet gained much of a foothold among philosophers of science. For a discussion, see Collins and Cox, "Recovering Relativity," and Lefebvre, "What Is the Historical Past?"

32. This event is well known to historians of biology. For a discussion featuring the philosophical aspects, see Brannigan, "The Law Valid for Pisum," or the similar treatment in his *The Social Basis of Scientific Discoveries*. Brannigan's book is a stimulating discussion of the social aspects of scientific discovery.

33. Holton, "Can Science Be Measured?," p. 43.

34. Kuhn, "The Halt and the Blind."

35. Ibid., p. 183. Kuhn observes this point in practice in that no mention of the apparatus of *The Stucture of Scientific Revolutions* occurs in *Black-Body Theory*, which might have been written without awareness of the distinction between normal and revolutionary science, and the role of paradigms in the former.

Chapter 4. Scientific Facts and Scientific Theories

1. *Negotiation* is explained in Bloor, *Knowledge and Social Imagery*, pp. 117–140. The strong program with which it is associated is developed in this book and in Barnes, *Scientific Knowledge*. See the relevant discussion of the strong program in Woolgar, "Interests and Explanation."

2. Ravetz, *Scientific Knowledge and Its Social Problems*, pp. 69–240.

3. Holton, *The Scientific Imagination*, pp. 25–83.

4. Fleck, *Genesis and Development*, p. 118.

5. Ibid., p. 72.

6. Ibid., p. 81.

7. Latour and Woolgar, *Laboratory Life*, especially pp. 105–150.

8. Ibid., p. 129.

9. Ibid., p. 130.

10. Ibid., p. 134.

11. Ibid., p. 138.

12. Elkana, et al., eds., *Towards a Metric*, frontispiece, with appropriate woodcut.

13. Hanson, *Patterns of Discovery*.

14. Popper, *Objective Knowledge*, especially chap. 4.

15. As is reported in many histories of astronomy, some early observers

saw Saturn as though it were a sphere with two handles attached, like those on a teacup. It was some time before these early guesses were eliminated by considerations of theory and better evidence.

16. There is a good example in Fleck, *Genesis and Development*, 135.

17. Clavelin, *The Natural Philosophy of Galileo*, pp. 393–403.

18. Hanson, *Patterns of Discovery*, pp. 100–102.

19. Collins, "The TEA set," and Collins and Harrison, "Building a TEA Laser."

20. Collins, "The Seven Sexes."

21. Lakatos, *Proofs and Refutations*.

22. Rosenberg, *Perspectives on Technology*, pp. 269–270.

23. Bloor, *Knowledge and Social Imagery*, and Klein, *Greek Mathemathical Thought*.

24. The foundational disputes in set theory, on the other hand, are a sign that this domain hasn't been settled.

25. Kuhn, *The Essential Tension*, pp. 340–351.

26. Gablik, *Progress in Art*.

27. See Steinberg, *Other Criteria*, especially pp. 55–93.

28. Kuhn, "The Halt and the Blind."

29. Fisher, "The Death of a Mathematical Theory," and "The Last Invariant Theories."

30. Fisher, "The Death of a Mathematical Theory," p. 138.

31. Kuhn suggests in "The Halt and the Blind" that theories are replaced only in science; Fisher makes a similar point in "The Last Invariant Theories," p. 222. The greater precision of border of domain may assist preservation in mathematics, and make it more like the overt cases of preservation in science, like the longevity associated with Newton's theory.

32. Latour and Woolgar, *Laboratory Life*, pp. 176–180.

33. Zahar, "Einstein, Meyerson, and the Role of Mathematics," pp. 2–8.

34. See Ackermann, "Confirmatory Models of Theories," and the discussion of Boltzmann in Bellone, *A World on Paper*.

35. The prediction suggested that Neptune could be found within a certain distance from a calculated point. Such predictions are part of the history of all sciences.

36. Latour and Woolgar, *Laboratory Life*. See also the discussion of language in Gilbert and Mulkay, "Warranting Scientific Belief."

37. Gay, "Radicals and Types," pp. 14–15.

38. Fleck, *Genesis and Development*.

39. Heisenberg, "Der Begriff 'abgeschlossene Theorie.' "

40. Ibid.

41. Hanson, *Patterns of Discovery*.

42. Fleck, *Genesis and Development*, p. 30.

43. The unified science hypothesis is the view that all of science is potentially reducible to physics, according to rules that would translate any state-

ment in any science into a statement of physical theory or observation, or both, for this purpose.

44. Plantinga, *God, Freedom, and Evil,* and Ackermann, "An Alternative Free Will Defense."

45. Goodman, *Fact, Fiction, and Forecast.*

46. Themata are discussed in Holton, *Thematic Origins.* For examples of themata in scientific history, see also Gruber, "Darwin's 'Tree of Nature,' " Wise, "The Mutual Embrace," and Finocchiaro, "Scientific Discoveries."

47. That metaphor and analogy are important is widely recognized, but analysis has proved difficult. For an interesting application, see Darden, "Theory Construction."

48. Hattiangadi, "The Vanishing Context of Discovery."

49. Gruber, "Darwin's 'Tree of Nature.' "

50. Wise, "The Mutual Embrace."

51. Miller, "Visualization Lost and Regained," and Hanson, *Patterns of Discovery.*

52. Neptune was predicted as an additional planet. For the positron, a new particle, see Hanson, *The Concept of the Positron.*

53. See the discussion of cosmological principles in North, *The Measure of the Universe.*

54. Barber and Fox, "The Case of the Floppy-Eared Rabbits."

55. Carter, "The Germ Theory."

56. Webster, "The Discovery of Boyle's Law."

57. See ibid., for details.

58. Gay, "Radicals and Types." This is a particularly stimulating essay in terms of its philosophical implications.

59. White, Sullivan, and Barboni, "The Interdependence of Theory and Experiment."

60. Ibid.

61. See Gay, "Radicals and Types," pp. 25–26, as well as White, Sullivan, and Barboni, "The Interdepencence of Theory and Experiment," p. 323.

Appendix. The Human Sciences

1. The human sciences are here opposed to the natural sciences, rather than opposing the behavioral sciences or the social sciences to the natural sciences. These latter oppositions often involve the introduction of a methodological point of view. We do not which to exclude history, textual criticism, and related activities as outside the scope of scientific activity. We also avoid the assumption that a methodological break occurs when man is the object of study. In opposing the human sciences to the natural sciences, we are trying to hold various possibilities open during the preliminary discussion. See Goldmann, *The Human Sciences and Philosophy,* for a related point of view.

2. The relationships between the natural and the social sciences have been

examined in the special case of physics and sociology in Chapter 2 in the section on scientific disciplines. It was argued there that substantive difference between physics and sociology could not be found through sociological means. This point is being reconsidered here now that our general account of theory and experiment has been concluded.

3. Values cannot be used (in general) to differentiate human from natural sciences, since the craft nature of scientific work means that valuations are inextricably involved in basic experimental research. But human values are also involved in a different sense, even in physics. The value of human understanding and control of physical nature is presupposed as something that is in the interests of all human beings.

4. These differing relationships may or may not be thought subject to further analysis. For an analytic treatment, see von Wright, *Explanation and Understanding*, and the discussion in Manninen and Tuomela, *Essays on 'Explanation and Understanding.'* Van Parijs, *Evolutionary Explanation*, also presents an interesting analysis of social science explanatory models. Like Popper's attempt to introduce the rationality principle, these analytic approaches have the practical consequence of assimilating parts of the human science domain to the natural sciences, in effect extending the analytical boundaries of natural science methodology by favoring areas of the social sciences that are closest to the natural sciences, given the methodologies in question. More hermeneutical accounts motivated by an intuition of decisive break are discussed in Dreyfus, "Holism and Hermeneutics," Taylor, "Understanding in Human Science," and Rorty, "A Reply to Dreyfus and Taylor." Within the traditional accounts of methodology, the distinction can always be drawn. See Lammers, "Mono- and Poly-Paradigmatic Developments," for the derivative Kuhnian claim that the social sciences are poly-paradigmatic and the natural sciences mono-paradigmatic, which seems to ignore the presence of real controversy in the natural sciences.

5. Thiel, *Grundlagenkrise und Grundlagenstreit*, provides interesting documentation that foundational disputes in the social sciences threaten data back to the beginning of the disciplines. In the natural sciences and mathematics, this retrogression is blocked by the last secure data domain, meaning that less damage accrues from foundational upheavals. Instruments are not available in the social sciences to break the connection between theory and fact.

6. The theories are too large in scope in the human sciences to suggest suitable instruments to embody their perceptions of reality. Theory thus reaches to fact, and colors it too strongly for the purposes of dialectical progress. Under these circumstances, the call for more data or more theory, or more middle-level theory, within the present context is to ensure the status quo. This is also true of versions of Marxism that attempt to develop discourse independently of the constraint of fact, a development that can elaborate, but not criticize, its own starting assumptions no matter how corrosive its attack on alternatives. See Hindess, *Philosophy and Methodology* for an example.

For discussions of relationship of theory to data in the social sciences, see Hindess, *The Use of Official Statistics*, and Turner, *Ethnomethodology*. Menzies, *Sociological Theory in Use*, deserves special study. Menzies shows that the attempt to circumscribe empirical research by intuitive theorizing results in research data that usually bear little specific imprint of the restricted theories, with the result that the data cannot sharply confirm or falsify the theoretician's claims. This is the other side of the inability to determine sharp data domains within the human sciences.

7. I am indebted for this point to a conversation with Margaret Nash.

8. The location of the boundary between the individual and society is at least partly a matter of convention. Clearly psychological structure is in some sense a process of acculturation, and social structure may in turn be jarred by nonassimilated aspects of individuality. Various attempts to synthesize Freud and Marx, such as those of the Frankfurt School, confront this situation, but usually with quite differing insights.

9. This is obscured by treatments that attempt to associate neoclassical and Marxist ideas with philosophies of the natural sciences, instead of with the natural sciences directly. Although neoclassical economists often defend themselves methodologically by positivist arguments, a refutation of positivism is not a refutation of classical theory. See Hollis and Nell, *Rational Economic Man*, for the association, and Ackermann, "Methodology and Economics," for a brief discussion. There is no methodological reason why Marxists can't be positivists in terms of economic methodology, unless very coercive definitions are introduced to define the possible parameters.

10. Durkheim's attempt to find an empirical realm of social fact should also be mentioned as an attempt to create a data domain for the social sciences. Habermas has recently attempted to mediate between Adorno's holism and analytical empiricism by finding a distinction between cognitive interests and sciences that pursue these interests. See Habermas, *Knowledge and Human Interests*. This permits a distinction between natural science methodology and hermeneutics to be drawn, to which is added a notion of critical science based on an analytic theory of communicative competence. See Habermas, *Communication and the Evolution of Society*, for a sketch of the project. Habermas must distinguish himself from natural science methodological imperialism (see the debate in Habermas and Luhmann, *Theorie der Gesellschaft oder Sozialtechnologie*) as well as from the holism of critical theory. The price that he has paid is to be seen from each camp as belonging to the other, and to some extent he therefore pays the price of mediation earlier paid by Weber. This situation is too delicate for detailed treatment here.

11. An explicit relationship between neoclassical theory and physical theory is described in Friedman, "The Methodology of Positive Economics." It is completely defensible from the methodological point of view. See the discussion in Diesing, *Patterns of Discovery*, Nagel, "Assumptions in Economic Theory," Rosenberg, *Microeconomic Laws*, and the overview in Ackermann, "Methodology and Economics." Becker, *The Economic Approach*, illustrates

analogies between extending the scope of theory in natural science and the extension of neoclassical theory.

12. See Gide and Rist, *A History of Economic Doctrines*, for the relevant history of early failure of prediction, and attempts to correct the problem.

13. Kornai, *Anti-Equilibrium*.

14. Leibenstein, *Beyond Economic Man*.

15. See Ackermann, "Methodology and Economics."

16. See Sraffa, *Production of Commodities*, and discussion of its significance in Pasinetti, *Lectures on the Theory of Production*, and Steedman, *Marx after Sraffa*.

17. For tame versions of *verstehen* as a mere heuristic, see Abel, "The Operation Called Verstehen," and Truzzi, *Verstehen*.

18. Weber, of course, assumed a questionable view of the natural sciences, but assumed that the methodology of *ideal types* could mediate the territory between precise refutation in the natural sciences and the problematic nature of data in the social sciences. See Giddens, *Positivism and Sociology*, p. 26, where Weber is quoted this way:

> If a hypothetical 'law of nature' definitely fails in only one instance, it collapses once and for all as a hypothesis. The ideal-typical constructions of political economy, however, do not—correctly understood—claim general validity, whereas a 'law of nature' must make this claim if it is not to lose its significance. A so-called 'empirical' law in the end is an empirically validated rule, having a problematic causal interpretation. A teleological schema of rational action, on the other hand, is an interpretation with a problematical empirical validation: the two are thus polar opposites. But both schemata are 'ideal typical' conceptual constructions. . . .

Weber's conception here has an obvious relationship to our conception of the relationship as based on entry points of understanding.

19. See Stammer, *Max Weber and Sociology Today*, pp. 27–88.

20. The hold of this distinction is clearly seen in such sources as Meissner and Wold, "The Foundations of Science," especially p. 127.

21. Held, *Introduction to Critical Theory*, and Jay, *The Dialectical Imagination*, provide general introductory accounts of critical theory.

22. Stammer, *Max Weber and Sociology Today*, p. 66.

23. The major documents of the *Positivismusstreit* are collected and translated in Adorno et al., *The Positivist Dispute*.

24. Ibid., pp. 288–289.

25. Ibid., p. 106.

26. Popper, *Conjectures and Refutations*, p. 342. See Moon, "The Logic of Political Inquiry," p. 157, and Ackermann, *The Philosophy of Karl Popper*, p. 171, for examples of Popperian generalizations in the social sciences.

27. Discussion of these matters begins with Jacobs, *The Death and Life of Great American Cities*.

28. The general thesis is expressed in Adorno, *Negative Dialectics*. Ador-

no's work is, however, concerned with general historical theses, as in Adorno and Horkheimer, *Dialectic of Enlightenment*, as well as with empirical data. Adorno's encounter with empirical data and his position that they can only be grasped through careful theoretical reflection can be located in Adorno et al., *The Authoritarian Personality*, Adorno, *Quasi una Fantasia*, "Scientific Experiences," "The Stars Down to Earth," and "Über Jazz." See the comments of an empiricist on Adorno's work in Lazarsfeld, "An Episode."

29. Adorno et al., *The Positivist Dispute*, p. 105.
30. Ibid., p. 106.
31. Ibid., pp. 112–113.
32. Ibid., p. 77.
33. Ibid., p. 85.
34. See the important survey in Menzies, *Sociological Theory in Use*.
35. See Adorno et al., *The Positivist Dispute*, pp. 131–257.

BIBLIOGRAPHY

Abel, Theodore. "The Operation Called Verstehen." *American Journal of Sociology* 54 (1948): 211–218.

Ackermann, Robert J. "An Alternative Free Will Defense." *Religious Studies* 18 (1982): 365–372.

———. "Confirmatory Models of Theories." *British Journal for the Philosophy of Science* 14 (1963): 89–105.

———. "Inductive Simplicity." *Philosophy of Science* 28 (1961): 152–161.

———. "Methodology and Economics." *The Philosophical Forum* 14 (1983): 389–402.

———. *The Philosophy of Karl Popper*. Amherst: University of Massachusetts Press, 1976.

———. *The Philosophy of Science*. New York: Pegasus, 1970.

Adorno, Theodore W. *Negative Dialectics*. Translated by E. B. Ashton. New York: Seabury Press, 1979.

———. *Quasi una Fantasia*. Frankfurt am Main: Suhrkamp, 1963.

———. "Scientific Experiences of a European Scholar in America." Translated by D. Fleming. In D. Fleming and B. Bailyn, eds., *The Intellectual Migration: Europe and America 1930–1960*, pp. 338–370. Cambridge, Mass.: Harvard University Press, 1969.

———. "The Stars Down to Earth." *Telos* 19 (1974): 13–90.

———. "Über Jazz." *Zeitschrift für Sozialforschung* 5 (1936): 235–257.

Adorno, Theordore W., et al. *The Authoritarian Personality*. New York: W. W. Norton and Co., 1969.

Adorno, Theodore W., et al. *The Positivist Dispute in German Sociology*. Translated by G. Adey and B. Frisby. New York: Harper and Row, 1976.

Adorno, Theodore W., and Horkheimer, Max. *Dialectic of Enlightenment*. Translated by J. Cumming. New York: Herder and Herder, 1969.

Baldamus, W. "Relevanz und Trivialität in der soziologischen Forschung." *Zeitschrift für Soziologie* 2 (1973): 2–20.

Ball, W. W. Rouse, and Coxeter, H.M.S. *Mathematical Recreations and Essays*. 12th ed. Toronto: University of Toronto Press, 1974.

Barber, Bernard, and Fox, Renee C. "The Case of the Floppy-Eared Rabbits: An Instance of Serendipity Gained and Serendipity Lost." *American Journal of Sociology* 64 (1958): 128–136.

Barnes, Barry. *Scientific Knowledge and Sociological Theory*. London: Routledge and Kegan Paul, 1974.

Barnes, Barry, and Dolby, R. G. "The Scientific Ethos: A Deviant Viewpoint." *European Journal of Sociology* 2 (1970): 3–25.

Barnes, Barry, and Edge, David, eds. *Science in Context*. Cambridge, Mass.: MIT Press, 1982.

Bastin, Ted, ed. *Quantum Theory and Beyond*. Cambridge: Cambridge University Press, 1978.

Bauman, Zygmunt. *Hermeneutics and Social Science*. New York: Columbia University Press, 1978.

Bazerman, Charles. "What Written Knowledge Does: Three Examples of Academic Discourse." *Philosophy of the Social Sciences* 11 (1981): 361–387.

Becker, Gary S. *The Economic Approach to Human Behavior*. Chicago: University of Chicago Press, 1976.

Bellone, Enrico. *A World on Paper*. Cambridge, Mass.: MIT Press, 1980.

Beyer, Janice M., and Stevens, John M. "Forschungsaktivität und Produktivität." In Nico, Stehr and Rene König, eds., *Wissenschaftssoziologie, Sonderheft* 18, *Kölner Zeitschrift für Soziologie und Sozialpsychologie* (1975): 349–374.

Bhasker, Roy. *A Realist View of Science*. Sussex: Harvester Press, 1978.

Bloor, David. *Knowledge and Social Imagery*. London: Routledge and Kegan Paul. 1976.

Blume, Stuart S., and Sinclair, Ruth. "Aspects of the Structure of a Scientific Discipline." In R. Whitley, ed., *Social Processes of Scientific Development*, pp. 224–241. London: Routledge and Kegan Paul, 1974.

Bohm, David. *Causality and Chance in Modern Physics*. New York: Harper and Row, 1961.

Bourdieu, Pierre. "The Specificity of the Scientific Field and the Social Conditions of the Progress of Reason." *Social Science Information* 14 (1975): 19–47.

Brannigan, Augustine. "The Law Valid for Pisum." Paper presented at the first Leonard Conference, Reno, Nevada, Fall 1978. See Augustine Brannigan, *The Social Basis of Scientific Discoveries*, chap. 6.

———. *The Social Basis of Scientific Discoveries*. Cambridge: Cambridge University Press, 1981.

Bressan, Aldo. *A General Interpreted Modal Calculus*. New Haven: Yale University Press, 1972.

Brittan, Gordon, G., Jr., and Lambert, Karel. *An Introduction to the Philosophy of Science.* Englewood Cliffs, N.J.: Prentice-Hall, 1970.

Cargile, James T. "Newcomb's Paradox." *British Journal for the Philosophy of Science* 26 (1975): 234–239.

Carnap, Rudolf. *Logical Foundations of Probability.* 2d ed. Chicago: University of Chicago Press, 1962.

———. "The Methodological Character of Theoretical Concepts." In H. Feigl and M. Scriven, eds., *Minnesota Studies in the Philosophy of Science* vol. 1, pp. 38–77. Minneapolis: University of Minnesota Press, 1956.

Carnap, Rudolf, and Jeffrey, Richard C. *Studies in Inductive Logic and Probability*, vol. 1. Berkeley: University of California Press, 1971.

Carter, K. Codell. "The Germ Theory, Beriberi, and the Deficiency Theory of Disease." *Medical History* 21 (1977): 119–136.

Clavelin, Maurice. *The Natural Philosophy of Galileo.* Translated by A. J. Pomerans. Cambridge, Mass.: MIT Press, 1974.

Cole, Stephen; Cole, Jonathan; and Dietrich, Lorraine. "Measuring the Cognitive State of Scientific Disciplines," In Y. Elkana, J. Lederberg, R. K. Merton, A. Thackray, and H. Zuckerman et al., eds., *Towards a Metric of Science*, pp. 209-251. New York: John Wiley and Sons, 1978.

Collins, H. M. "The Seven Sexes: A Study in the Sociology of a Phenomenon, or the Replication of Experiment in Physics." *Sociology* 9 (1975): 205–224.

———. "The TEA Set: Tacit Knowledge and Scientific Networks." *Science Studies* 4 (1974): 165–186.

Collins, H. M., and Cox, Graham. "Recovering Relativity: Did Prophecy Fail?" *Social Studies of Science* 6 (1976): 423–444.

Collins, H. M., and Harrison, R. G. "Building a TEA Laser: The Caprices of Communication." *Social Studies of Science* 5 (1975): 441–450.

Collins, H. M., and Pinch, Trevor J. "The Construction of the Paranormal: Nothing Unscientific Is Happening." In Roy Wallis, ed., *On the Margins of Science*, pp. 237–270. Keele: University of Keele, 1979.

———. *The Social Construction of Extraordinary Science.* London: Routledge and Kegan Paul, 1982.

Conant, James B. "The Overthrow of the Phlogiston Theory." In James B. Conant, ed., *Case Histories in Experimental Science*, vol. 1, pp. 65–116. Cambridge, Mass.: Harvard University Press, 1957.

Crane, Diana, "An Exploratory Study of Kuhnian Paradigms in Theoretical High Energy Physics." *Social Studies of Science* 10 (1980): 23–54.

Darden, Lindley. "Theory Construction in Genetics." In Thomas Nickles, *Scientific Discovery: Case Studies*, pp. 151–170.

Diesing, Paul. *Patterns of Discovery in the Social Sciences.* Chicago: Aldine Publishing Co., 1971.

Dreyfus, Hubert L. "Holism and Hermeneutics." *Review of Metaphysics* 34 (1980): 3–23.

Eberle, R.; Kaplan, D.; and Montague, R. "Hempel and Oppenheim on Explanation." *Philosophy of Science* 28 (1961): 418–428.

Ehrlich, Paul R., and Feldman, S. Shirley. *The Race Bomb.* New York: Ballantine Books, 1977.

Elkana, Y.; Lederberg, J.; Merton, R. K.; Thackray, A.; and Zuckerman, H., et al., eds. *Towards a Metric of Science.* New York: John Wiley and Sons, 1978.

Elsasser, Walter M. *Atom and Organism.* Princeton: Princeton University Press, 1966.

Evenson, Robert E.; Waggoner, Paul E.; and Ruttan, Vernon W. "Economic Benefits from Research: An Example from Agriculture." *Science* 205 (1979): 1101–1107.

Ferguson, Eugene S. "The Mind's Eye: Nonverbal Thought in Technology." *Science* 197 (1977): 827–836.

Feyerabend, Paul. *Against Method.* London: New Left Books, 1975.

Finacchiaro, Maurice F. "Scientific Discoveries as Growth of Understanding: The Case of Newton's Gravity." In Thomas Nickles, *Scientific Discovery: Case Studies*, pp. 235–255.

Fisher, Charles S. "The Death of a Mathematical Theory." *Archive for History of Exact Sciences* 3 (1966): 137–159.

———. "The Last Invariant Theories." *European Journal of Sociology* 8 (1967): 216–244.

Fleck, Ludwik. *Genesis and Development of a Scientific Fact.* Translated by F. Bradley and T. J. Trenn. Chicago: University of Chicago Press, 1979.

Friedman, Milton, "The Methodology of Positive Economics." In Milton Friedman, *Essays in Positive Economics.* Chicago: University of Chicago Press, 1953.

Gablik, Suzi. *Progress in Art.* London: Thames and Hudson, 1976.

Garfield, Eugene; Malin, Morton V.; and Small, Henry. "Citation Data as Science Indicators." In Y. Elkana, J. Lederberg, R. K. Merton, A. Thackray, H. Zuckerman, et al., eds., *Towards a Metric of Science*, pp. 179–207.

Gay, Hannah. "Radicals and Types: A Critical Comparison of the

Methodologies of Popper and Lakatos and Their Use in the Reconstructions of Some Nineteenth-Century Chemistry." *Studies in History and Philosophy of Science* 6 (1976): 1–51.

Giddens, Anthony. *Positivism and Sociology*. London: Heinemann, 1974.

Gide, Charles, and Rist, Charles. *A History of Economic Doctrines*. 2d ed. London: Macmillan Publishing Co., 1948.

Gilbert, G. N. "The Development of Science and Scientific Knowledge: The Case of Radar Meteor Research." In G. Lemaine, R. MacLeod, M. Mulkay, P. Weingart, eds., *Perspectives on the Emergence of Scientific Disciplines*, pp. 187–204. The Hague: Mouton, 1976.

Gilbert, G. N., and Mulkay, Michael. "Warranting Scientific Belief." *Social Studies of Science* 12 (1982): 383–408.

Gillespie, Brendan; Eve, Dave; and Johnston, Ron. "Carcinogenic Risk Assessment in the USA and UK: The Case of Aldrin/Dieldrin." In Barry Barnes and David Edge, eds., *Science in Context*, pp. 303–335.

Goldmann, Lucien. *The Human Sciences and Philosophy*. London: Grossman Publishers, 1969.

Goodman, Nelson. *Fact, Fiction, and Forecast*. 2d ed. Indianapolis: Bobbs-Merrill Company, 1965.

———. *Languages of Art*. Indianapolis: Bobbs-Merrill Company, 1968.

———. *Ways of Worldmaking*. Indianapolis: Hackett Publishing Company, 1978.

Griffith, Belver C., and Mullins, Nicholas C. "Coherent Social Groups in Scientific Change." *Science* 177 (1972): 959–964.

Griffith, Belver C., and Small, Henry G. "The Structure of Scientific Literatures." *Science Studies* 4 (1974): 17–40, 339–365.

Gruber, Howard. "Darwin's 'Tree of Nature' and Other Images of Wide Scope." In J. Wechsler *Aesthetics in Science*, pp. 121–140. Cambridge, Mass.: MIT Press, 1979.

Habermas, Jürgen. *Communication and the Evolution of Society*. Translated by T. McCarthy. Boston: Beacon Press, 1976.

———. *Knowledge and Human Interests*. Translated by J. Shapiro. Boston: Beacon Press, 1971.

Habermas, Jürgen, and Luhmann, Niklas. *Theorie der Gesellschaft oder Sozialtechnologie*. Frankfurt am Main: Suhrkamp, 1976.

Hacking, Ian. Review of *The Structure of Scientific Revolutions*, by Thomas Kuhn. *History and Theory* 17 (1978): 233–236.

Hannaway, Owen. *The Chemists and the Word*. Baltimore: Johns Hopkins University Press, 1975.

Hanson, Norwood Russell. *The Concept of the Positron*. Cambridge: Cambridge University Press, 1963.

———. *Patterns of Discovery*. Cambridge: Cambridge University Press, 1958.

Hattiangadi, J. N. "The Vanishing Context of Discovery: Newton's Discovery of Gravity." In Thomas Nickles, *Scientific Discovery, Logic, and Rationality*, pp. 257–265.

Heisenberg, Werner, "Der Begriff 'abgeschlossene Theorie' in Der modernen Naturwissenschaft." In W. Heisenberg, *Schritte Über Grenzen*, pp. 87–94. Munich: R. Piper and Co., 1971.

Held, David. *Introduction to Critical Theory*. Berkeley: University of California Press, 1980.

Hempel, Carl G. *Aspects of Scientific Explanation*. New York: Free Press, 1965.

———. "Deductive-Nomological vs. Statistical Explanation." In H. Feigl and G. Maxwell, eds., *Minnesota Studies in the Philosophy of Science*, vol. 3, pp. 98–169. Minneapolis: University of Minnesota Press, 1962.

Hindess, Barry. *Philosophy and Methodology in the Social Sciences*. Atlantic Highlands, New Jersey: Humanities Press, 1977.

———. *The Use of Official Statistics in Sociology: A Critique of Positivism and Ethnomethodology*. London: Macmillan, 1973.

Hollis, M., and Nell, E. *Rational Economic Man: A Philosophical Critique of Neo-Classical Economics*. Cambridge: Cambridge University Press, 1975.

Holton, Gerald. "Can Science Be Measured?" In Y. Elkana, J. Lederberg, R. K. Merton, A. Thackray, H. Zuckerman, et al., eds., *Towards a Metric of Science*, pp. 39–68.

———. *The Scientific Imagination: Case Studies*. Cambridge: Cambridge University Press, 1978.

———. *Thematic Origins of Scientific Thought*. Cambridge, Mass.: Harvard University Press, 1973.

Hounshell, David A. "Edison and the Pure Science Ideal in Nineteenth-Century America." *Science* 207 (1980): 612–617.

Howson, Colin, ed. *Method and Appraisal in the Physical Sciences: The Critical Background to Modern Science, 1800–1905*. Cambridge: Cambridge University Press.

Hull, David L.; Tessner, Peter D.; and Diamond, Arthur M. "Planck's Principle: Do Younger Scientists Accept New Scientific Ideas with Greater Alacrity Than Older Scientists?" *Science* 202 (1978): 717–723.

Ivins, William M. *Prints and Visual Communication.* Cambridge: MIT Press, 1953.

Jacobs, Jane. *The Death and Life of Great American Cities.* New York: Random House, 1961.

Jay, Martin. *The Dialectical Imagination.* Boston: Little, Brown, and Company, 1973.

Klein, J. *Greek Mathematical Thought and the Origin of Algebra.* Cambridge, Mass.: MIT Press, 1968.

Knorr, Karin. "The Nature of Scientific Consensus and the Case of the Social Sciences." In K. Knorr, H. Strasser, and H. G. Zilian, eds., *Determinants and Controls of Scientific Development*, pp. 227–256. Dordrecht: D. Reidel Publishing Co., 1975.

Knorr-Cetina, Karin D. "Scientific Communities or Transepistemic Arenas of Research? A Critique of Quasi-Economic Models of Science." *Social Studies of Science* 12 (1982): 101–130.

———. "Social and Scientific Method, or What Do We Make of the Distinction between the Natural and the Social Sciences?" *Philosophy of the Social Sciences* 11 (1981): 335–359.

Koester, David; Sullivan, Daniel; and White, D. Hywel. "Theory Selection in Particle Physics: A Quantitative Case Study of the Evolution of Weak-Electromagnetic Unification Theory." *Social Studies of Science* 12 (1982): 73–100.

Kordig, Carl R. *The Justification of Scientific Change.* Dordrecht: D. Reidel Publishing Co., 1971.

Kornai, Janos. *Anti-Equilibrium: On Economics Systems Theory and the Tasks of Research.* Amsterdam: North-Holland Publishing Co., 1971.

Krüger, Lorenz. "Die systematische Bedeutung wissenschaftlicher Revolutionen, Pro und Contra Thomas Kuhn." In W. Diedrich, ed., *Theorien der Wissenschaftsgeschichte*, pp. 210–246. Frankfurt am Main: Suhrkamp, 1978.

———. "Revolutionen und Kontinuität der Erfahrung." *Neue Hefte für Philosophie* 6/7 (1974): 1–26.

Kuhn, Thomas S. *Black-Body Theory and the Quantum Discontinuity, 1894–1912.* Oxford: Clarendon Press, 1978.

———. *The Essential Tension.* Chicago: University of Chicago Press, 1977.

———. "The Halt and the Blind." Review of *Method and Appraisal in the Physical Sciences*, edited by Colin Howson. *British Journal for the Philosophy of Science* 31 (1980): 181–192.

———. *The Structure of Scientific Revolutions.* Chicago: University of Chicago Press, 1970.

Lakatos, Imre. *Philosophical Papers*, vols. 1 and 2. Cambridge: Cambridge University Press, 1978.

———. *Proofs and Refutations*. Cambridge: Cambridge University Press, 1976.

Lakatos, Imre, and Musgrave, Alan, eds. *Criticism and the Growth of Knowledge*. Cambridge: Cambridge University Press, 1970.

Lammers, Cornelius J. "Mono- and Poly-Paradigmatic Developments in Natural and Social Sciences." In R. Whitley, ed., *Social Processes of Scientific Development*, pp. 123–147. London: Routledge and Kegan Paul, 1974.

Latour, Bruno, and Woolgar, Steve. *Laboratory Life*. Sage Library of Social Research, vol. 80. London: Sage Publications, 1979.

Laudan, Larry. *Progress and Its Problems*. Berkeley: University of California Press, 1977.

Lazarsfeld, Paul. "An Episode in the History of Social Research: A Memoir." In D. Fleming and B. Bailyn, eds., *The Intellectual Migration: Europe and America, 1930–1960*, pp. 322–337. Cambridge, Mass.: Harvard University Press, 1969.

LeCourt, Dominique. *Marxism and Epistemology*. Translated by B. Brewster. London: New Left Books, 1975.

Lefebvre, Henri. "What Is the Historical Past?" *New Left Review* 90 (1975): 27–34.

Leibenstein, Harvey. *Beyond Economic Man*. Cambridge, Mass.: Harvard University Press, 1976.

Levi, Isaac. *The Enterprise of Knowledge*. Cambridge, Mass.: MIT Press, 1980.

———. "Newcomb's Many Problems." *Theory and Decision* 6 (1975): 161–175.

McCarthy, Thomas. *The Critical Theory of Jürgen Habermas*, Cambridge, Mass.: MIT Press, 1978.

MacKenzie, Donald A. Statistics in Britain, 1865–1930: The Social Construction of Scientific Knowledge. Edinburg: Edinburg University Press, 1981.

Manninen, Juha, and Tuomela, Raimo. *Essays on 'Explanation and Understanding': Studies in the Foundations of Humanities and Social Sciences*. Dordrecht: D. Reidel Publishing Co., 1976.

Martin, Brian. *The Bias of Science*. Canberra: Society for Social Responsibility in Science, 1979.

Martins, Herminio. "The Kuhnian 'Revolution' and Its Implications for Sociology." In T. J. Nossiter, A. H. Hanson, and S. Rokkan, eds., *Imagination and Precision in the Social Sciences*, pp. 13–58. London: Faber and Faber, 1972.

Meissner, W., and Wold, H. "The Foundations of Science in Cognitive Mini-Models." In W. Leinfellner and E. Köhler, eds., *Developments in the Methodology of Social Science*, pp. 111–146.

Menzies, Ken. *Sociological Theory in Use*. London: Routledge and Kegan Paul, 1982.

Merton, Robert K. *The Sociology of Science*. Chicago: University of Chicago Press, 1973.

Miller, Arthur I. "Visualization Lost and Regained: The Genesis of the Quantum Theory in the Period 1913–1927." In J. Wechsler, *Aesthetics in Science*, pp. 73–102. Cambridge, Mass.: MIT Press, 1979.

Miller, David. "Popper's Qualitative Theory of Verisimilitude." *British Journal for the Philosophy of Science* 25 (1974): 166–177.

Mitroff, Ian. *The Subjective Side of Science*. Amsterdam: Elsevier, 1974.

Moon, J. Donald. "The Logic of Political Inquiry." In F. Greenstein and N. Polsky eds., *Political Science in Scope and Theory*, pp. 131–228. London: Addison-Wesley, 1975.

Mulkay, M. J. *The Social Process of Innovation*. London: Macmillan Publishing Co., 1972.

Mullins, Nicholas C. "A Sociological Theory of Scientific Revolution." In K. Knorr, H. Strasser, and H. G. Zilian, eds., *Determinants and Controls of Scientific Development*, pp. 185–203. Dordrecht: D. Reidel Publishing Co., 1975.

Nagel, Ernest. "Assumptions in Economic Theory." *American Economic Review* 53 (1963): 211–220.

Nickles, Thomas. Papers from the first Guy E. Leonard Memorial Conference. Vol. 1, *Scientific Discovery, Logic, and Rationality*. Vol. 2, *Scientific Discovery: Case Studies*. Dordrecht: D. Reidel Publishing Co., 1980.

North, J. D. *The Measure of the Universe*. Oxford: Clarendon Press, 1965.

Nowotny, Helga. "Controversies in Science: Remarks on the Different Modes of Production of Knowledge and Their Use." *Zeitschrift für Soziologie* 4 (1975): 34–45.

Nozick, Robert. "Newcomb's Problem and Two Principles of Choice." In N. Rescher, ed., *Essays in Honor of Carl G. Hempel*, pp. 114-146. Dordrecht: D. Reidel Publishing Co., 1969.

Olby, Robert. *The Path to the Double Helix*. Seattle: University of Washington Press, 1974.

Olin, Doris. "Newcomb's Problem: Further Investigations." *American Philosophical Quarterly* 13 (1976): 129–133.

Pasinetti, Luigi L. *Lectures on the Theory of Production*. New York: Columbia University Press, 1977.

Pfetsch, Frank R. "The 'Finalization' Debate in Germany: Some Comments and Explanations." *Social Studies of Science* 9 (1979): 115–124.

Pickering, Andrew. "Interests and Analogies." In Barry Barnes and David Edge, eds., *Science in Context*, pp. 125–146.

Pinch, Trevor J. "Normal Explanations of the Paranormal: The Demarcation Problem and Fraud in Parapsychology." *Social Studies of Science* 9 (1979): 329–348.

———. "What Does a Proof Do If It Does Not Prove?" In E. Mendelsohn, P. Weingart, and R. Whitley, eds., *The Social Production of Scientific Knowledge*, pp. 171–215. Dordrecht: D. Reidel Publishing Co., 1977.

Plantinga, Alvin C. *God, Freedom, and Evil*. Grand Rapids: Eerdmans Publishing Co., 1977.

Popper, Karl R. *Conjectures and Refutations: The Growth of Scientific Knowledge*. New York: Harper and Row, 1965.

———. *Objective Knowledge*. Oxford: Clarendon Press, 1972.

Price, D. J. *Science since Babylon*. New Haven: Yale University Press, 1957.

Ravetz, Jerome R. *Scientific Knowledge and Its Social Problems*. Oxford: Clarendon Press, 1971.

Ronchi, Vasco. *New Optics*. Florence: Olschki, 1971.

Rorty, Richard. "A Reply to Dreyfus and Taylor." *Review of Metaphysics* 34 (1980): 39–46.

Rosenberg, Alexander. *Microeconomic Laws: A Philosophical Analysis*. Pittsburgh: University of Pittsburgh Press, 1976.

Rosenberg, Nathan. *Perspectives on Technology*. Cambridge: Cambridge University Press, 1976.

Rosenkrantz, Roger D. *Inference, Method, and Decision*. Dordrecht: D. Reidel Publishing Co., 1977.

Rosnow, Ralph L. *Paradigms in Transition*. New York: Oxford University Press, 1980.

Scheffler, Israel. *The Anatomy of Inquiry*. New York: Alfred A. Knopf, 1963.

———. *Science and Subjectivity*. Indianapolis: Bobbs-Merrill Company, 1967.

Shapere, Dudley. "Meaning and Scientific Change." In R. Colodny, ed., *Mind and Cosmos*, pp. 41–85. Pittsburgh: University of Pittsburgh Press, 1966.

Shapin, Steven. "The Politics of Observation: Cerebral Anatomy and

Social Interests in the Edinburgh Phrenology Disputes." In Roy Wallis, ed., *On the Margins of Science*, pp. 139-178. Keele: University of Keele Press, 1979.

Skyrms, Brian. *Causal Necessity*. New Haven: Yale University Press, 1980.

Sraffa, Piero. *Production of Commodities by Means of Commodities*. Cambridge: Cambridge University Press, 1960.

Stammer, Otto. *Max Weber and Sociology Today*. New York: Harper and Row, 1971.

Steedman, Ian. *Marx after Sraffa*. London: New Left Books, 1977.

Stegmüller, Wolfgang. *Probleme und Resultate der Wissenschaftstheorie und analytischen Philosophie*. Berlin: Springer Verlag, 1973.

Steinberg, Leo. *Other Criteria*. London: Oxford University Press, 1972.

Stokes, T. D. "The Double Helix and the Warped Zipper—An Exemplary Tale." *Social Studies of Science* 12 (1982): 207–240.

Stove, David. *Popper and After*. New York: Pergamon Press, 1982.

Taylor, Charles. "Understanding in Human Science." *Review of Metaphysics* 34 (1980): 25–38.

Tenbruch, Friedrich H. "Der Fortschritt der Wissenschaft als Trivialisierungsprozess." In Nico Stehr and Rene König, eds., *Wissenschaftssoziologie, Sonderheft* 18, *Kölner Zeitschrift für Soziologie und Sozialpsychologie* (1975): 19–47.

Thiel, Christian. *Grundlagenkrise und Grundlagenstreit*. Meisenheim am Glam: Anton Hain, 1972.

Tichy, Pavel. "On Popper's Definition of Verisimilitude." *British Journal for the Philosophy of Science* 25 (1974): 155–160.

Tizard, Jack. Discussion of "Progress and Degeneration in the 'IQ Debate.'" by Peter Urbach. *British Journal for the Philosophy of Science* 27 (1976): 251–258.

Toulmin, S. E. "Crucial Experiments: Priestley and Lavoisier." *Journal of the History of Ideas* 18 (1957): 205–220.

Truzzi, Marcello. *Verstehen: Subjective Understanding in the Social Sciences*. Reading, Mass.: Addison-Wesley Publishing Co., 1974.

Turner, Roy. *Ethnomethodology*. Harmondsworth: Penguin Books, 1974.

Urbach, Peter. "Progress and Degeneration in the 'IQ Debate.'" *British Journal for the Philosophy of Science* 25 (1974): 99–135, 235–259.

Van Parijs, Philippe. *Evolutionary Explanation in the Social Sciences*. Totowa, N.J.: Rowman and Littlefield, 1981.

von Wright, G. H. *Explanation and Understanding*. Cornell: Cornell University Press, 1971.

Watson, James D. *The Double Helix: A Personal Account of the Discovery of the Structure of DNA*. New York: Atheneum Press, 1968.

Webster, C. "The Discovery of Boyle's Law, and the Concept of the Elasticity of Air in the Seventeenth Century." *Archive for History of Exact Sciences* 2 (1965): 441–502.

White, D. Hywel; Sullivan, Daniel; and Barboni, Edward J. "The Interdependence of Theory and Experiment in Revolutionary Science: The Case of Parity Violation." *Social Studies of Science* 9 (1979): 303–327.

Whitley, Richard. "Changes in the Social and Intellectual Organization of the Sciences." In E. Mendelsohn, P. Weingart, and R. Whitley, eds., *The Social Production of Scientific Knowledge*, pp. 143–169. Dordrecht: D. Reidel Publishing Co., 1977.

———. "Types of Science, Organizational Structure, and Patterns of Work in Research Laboratories in Different Scientific Fields." *Social Science Information* 17 (1978): 427–447.

Widdowson, H. G. *Explorations in Applied Linguistics*. Oxford: Oxford University Press, 1979.

Wise, M. Norton. "The Mutual Embrace of Electricity and Magnetism." *Science* 203 (1979): 1310–1318.

Woolgar, Steve. "Interests and Explanation in the Social Study of Science." (Discussion following.) *Social Studies of Science* 11 (1981): 365–394.

Wright, Peter W. G. "A Study in the Legitimisation of Knowlege: The 'Success' of Medicine and the 'Failure' of Astrology." In Roy Wallis, ed., *On the Margins of Science*, pp. 85–101. Keele: University of Keele, 1979.

Wynne, Brian. "Between Orthodoxy and Oblivion: The Normalization of Deviance in Science." In Roy Wallis, ed., *On the Margins of Science*, pp. 67–84. Keele: University of Keele, 1979.

———. "C. G. Barkla and the J Phenomenon: A Case Study in the Treatment of Deviance in Physics." *Social Studies of Science* 6 (1976): 307–347.

Yearley, Steven. "The Relationship between Epistemological and Sociological Cognitive Interests: Some Ambiguities Underlying the Use of Interest Theory in the Study of Scientific Knowlege." *Studies in the History and Philosophy of Science* 13 (1982): 353–388.

———. "Textual Persuasion: The Role of Social Accounting in the

Construction of Scientific Arguments." *Philosophy of the Social Sciences* 11 (1981): 409–435.

Zahar, Elie. "Einstein, Meyerson, and the Role of Mathematics in Physical Discovery." *British Journal for the Philosophy of Science* 31 (1980): 1–43.

Ziman, John. *The Force of Knowledge*. Cambridge: Cambridge University Press, 1976.

INDEX

Adorno, T. W., 171, 178-186
agricultural technology, 86-87
astrology, its relationship to medicine, 81-82
autonomy of science, 96-97

Bachelard, G., 98-100
Barkla, C. G., 78-79, 147
Bellone, E., 64
beriberi, 158-159
big science, its impact, 41-42
Bloor, D., 140
Bohm, D., 68
Bourdieu, P., 63
Boyle's law, discovery of, 159-162
Brake Principle, 11-12, 14

Cartesian epistemology, 6, 93-94, 112
cheating in science, 40-41
citation research, 52-53, 58
closed theories, 148
Collins, H. M., 134-136
Compton, A. H., 78
confirmation, 17, 22-26
controversy in science, 36-40, 62-68, 112
Crane, D., 45-46

data domains, 30-31, 64, 73, 125-164, 169-171
differentiation, as a countertendency to competition, 66-69

economics, methodology of, 170-174
Edison, T. A., his status as a scientist, 85-86
Ehrenhaft, F., 117-121, 136
Einstein, A., 51
empiricism, 7-10, 16-29, 34, 109-110
explanation, 17-22; Hempelian model of, 20-22; in contrast with understanding, 166-167

Feyerabend, P., 28-29
fiction, science and, 31-33
finalization debate, 95-96
Fisher, C. S., 142-143
Fleck, L., 121-122, 147-149
floppy-eared rabbits, 158
free will, 165-166

Galileo, 28, 32, 80, 134, 159
game theory, 5-6
Gay, H., 147, 163
Gilbert, G. N., 66
Goodman, N., 153-154
gravity waves, 135-136

Hanson, N. R., 35
Heisenberg, W., 148
Hempel, C. G., 20-22
Holton, G., 109, 117, 154
Hume, D., 24
Husserl, E., 13

ideal types, 174-175
ideology, and norms in science, 37-40
instrumentalism, 30
instruments, scientific, 34, 49-51, 57-61, 87-88, 128-136, 168-169
integers, 144
internal history of science, its adequacy, 89-96
invariant theory, 142-143
IQ controversy, 91-92

Kepler, J., 90-91
Klein, J., 140
Kuhn, T. S., 26-30, 42-55, 62-68, 94, 98, 110, 114, 137, 142-143, 163-164

Lakatos, I., 110, 114, 138-139, 147, 163
language, its correspondence to the world, 12-15, 30-31, 136-137, 143-148
language of science, 100-104, 136-137
laser development and operation, 135
Latour, B., 122-124, 144, 147
Lavoisier, A., 104-106
Levi, I., 5
logic, empirical content of, 25-26

Martin, B., 39
Marxist class theory, 172-174
mathematical knowledge, analysis of, 3-4, 13-14, 52, 138-146
medicine, its relationship to astrology, 81-82
Mendel, G., 108
Merton, R. K., 36-39
Millikan, R. A., 117-121, 136, 147
motivation in science, 39-42

neoclassical economics, contrasted with physics, 171-172
Neptune, discovery of, 145-146, 156
Newcomb's puzzle, 6
Newton, I., 8, 50, 136, 148-149, 155, 159

objectivity, science and, 30-31, 40-41, 128-129

paradigms, Kuhn's treatment of, 43-50, 54-55, 80-81, 141
parapsychology, 75-77
parity conservation, overthrow of, 162-163
Pasteur, L., 79-80
phenomenology, 13
phlogiston theory, 104-106
phrenology, the debate over, 92-93
Plato, 3-4, 13-14
polycentrism, 55-57
Popper, K. R., 6, 23, 29, 62, 94, 114, 126, 137, 163, 171, 178-186
positivism, 16-20, 28, 50, 77
positivist dispute, 178-186
postparadigmatic science, the possibility of, 95-96
Priestley, J., 104-106
problem solving, and progress in science, 72-73
progress in science, 34, 49-51, 53-54, 58, 70-72, 150-164
pure science, 83-88, 91-96

quantum theory, 49-51, 155-156

radical theory, 161-162
rationalism, 7-15, 28, 109-110
rationality, analysis of, 4-6
Ravetz, J. R., 38, 94, 113-117, 121
realism, 30, 116
reduction, 152-153
relationships between scientific disciplines, 57-62
relativity theory, 51
research groups, and the structure of science, 42-62
revolutions, as part of scientific history, 50-51, 54-55, 68-70
Ronchi, V., 100

science: as a search problem, 70-71; its relationship to technology, 84-87, 96
science policy, 94-95
scientific facts, instruments as a basis for, 34, 117-120
skepticism, 3
social structure of science, 6, 35-73
Sraffa, P., 172-173
sufficient reason, principle of, 90

technology, its relationship to science, 84-87, 96
telescope, as creating a data domain, 134
textual interpretation, science and, 33, 37, 62-63, 111, 141-142
themata, 154-155
theory-laden observation, 31-33
type theory, 161-162

understanding, contrasted with explanation, 166-167

valence theory, 161-162
verstehen, 174-177

Watson, J. D., 38
Weber, M., 174-177
Whitley, R., 41
Wittgenstein, L., 69
Woolgar, S., 122-124, 144, 147

LIBRARY OF CONGRESS CATALOGING IN PUBLICATION DATA

Ackermann, Robert John, 1933-
Data, instruments, and theory.

Bibliography: p.
Includes index.
1. Science—Philosophy. 2. Science—Social aspects. 3. Logic.
I. Title. II. Title: Dialectical approach to understanding science.
Q175.A269 1985 501 84-15938
ISBN 0-691-07296-5 (alk. paper)